Science Under Siege

Science Under Siege

The Myth of Objectivity in Scientific Research

Beth Savan

CBC Enterprises

Montréal · Toronto · New York · London

Published by CBC Enterprises/Les Entreprises Radio-Canada,
a division of the Canadian Broadcasting Corporation, P.O. Box 500,
Station A, Toronto, Ontario M5W 1E6.

Canadian Cataloguing in Publication Data

Savan, Beth
Science under siege

Bibliography: p.
Includes index.
ISBN 0-88794-336-5

1. Research. 2. Science - Methodology.
I. CBC Enterprises. II. Title.

Q180.A1S38 1988   507'.2   C88-093353-4

Editor: Ex Libris/Charis Wahl
Cover Design: Peter Maher
Electronic Typesetting: Allen Shechtman,   Silver Bullet Productions

Distributed to the trade by the Canadian Book Marketing Group Ltd.
Printed and bound in Canada by D.W. Friesen and Sons Limited

1 2 3 4 5 6 / 93 92 91 90 89 88

To my mother
Kathleen (Green) Savan
1910 – 1981

This book is based on a four-part series produced by Beth Savan for CBC Radio "Ideas" which was awarded the Canadian Science Writers' Association Science and Health Award for excellence in science journalism.

Contents

... a wise estimate of consequences is fused in the fires of that passionate belief which determines the consequences it believes in. The inspirations of the world have come in that way too: even strictly-measuring science could hardly have got on without that forecasting ardour which feels the agitations of discovery beforehand, and has faith in its preconception that surmounts many failures of experiment. And in relation to human motives and actions, passionate belief has a fuller efficacy. Here enthusiasm may have the validity of proof....

George Eliot, *Daniel Deronda*, 1876

INTRODUCTION

Dr. Savan's book is an important contribution to the social critique of contemporary science, both as an activity and as an establishment.

Critique in this context has to be understood as the critical illumination of the design and practice of research, reflecting on the researchers as well as the sponsors of research. The screen onto which these reflections are projected is the political reality of our own society.

Within our time science plays a pivotal role, comparable to the role of religion in the Middle Ages. In the contemporary setting science provides the knowledge of how to live, how to approach and solve problems, and how to understand the universe and its workings. In the Middle Ages religious teachings filled these interpretive social functions and supplied the culture's common base.

But whether science or religion is regarded as the wellspring of vital knowledge, the lay population has – or has had – no direct access to it. In the quest for essential information, ordinary citizens have usually depended on the mediation and interpretations of an ordained caste of special people.

Be they scientists or priests, the members of such castes have always been schooled together in the use of common methodologies, which in turn have frequently and intentionally been rendered opaque to outsiders. Trained to function in hierarchal structures that give direction and reward to their search – or research – the ordained élites develop and express common views. Their loyalties reside principally with the establishment that nourishes them.

It is instructive to consider more parallels between the contemporary scientific establishment and, particularly, the pre-Reformation Church of Western Europe – beyond the structural similarities of their "ordained castes".

As in contemporary societies, the providers and dispensers of vital knowledge of the Middle Ages became adjuncts to the power structures and were used by the rulers of the day. In hindsight it is astounding how slowly the Reformation came, how long it took to establish the validity of the individual conscience, the direct access to vital insights, "the priesthood of all believers", which challenged established religious practices. By the same token, the modern public has traditionally accepted scientific practices, just because they are "scientific".

I do not intend to give up the collective search for more and deeper knowledge, but like Beth Savan, I consider knowledge a common good, and not an instrument to advance special power and interest. In order to retrieve or develop genuine and open processes of finding and sharing new and old knowledge, it is essential to break through the wall of unquestioning acceptance by citizens of any and all applications of "science".

Thus, it should be considered neither heresy nor subversion to question the design of research, its priorities, its funding, or the conclusions drawn from such research. Nor can the procurers of "scientific information" be immune from scrutiny and accountability.

A broadly based and principled critique of science as it is structured and practised today is the essential first step toward a long overdue reformation – the Reformation of Science.

Ursula M. Franklin
University of Toronto
January 1988

PREFACE

This book is the result of my exploration during the past decade of how the scientist and society are reflected in scientific research results. My initial concern with scientific bias grew out of my own struggle, as a naive doctoral candidate in the field of insect ecology, to challenge an accepted ecological model of predator behaviour. I am grateful to Professor T.R.E. Southwood, the head of the department where I was studying, for his support and for introducing me to the intellectual and political realities of scientific life. From my encounter with scientific conservatism grew a more general interest in how scientific careers and dogmas are constructed and maintained, which, with the guidance and inspiration of Max Allen, coalesced into the CBC Radio series "Science and Deception", aired on *Ideas* in 1982.

Popular response to this series was enormous; hundreds of letters, some critical but most, supportive, poured into the CBC. Researchers told of their experiences of prejudice, fraud, and élitism; and policy-makers expressed interest in the implications of my analysis for science funding. Science teachers were keen on the alternative view of scientific research the program projected, and academic scientists were divided between those who welcomed my approach as a vindication of their own difficulties and those who criticized the series as unfair or unfounded. The widespread interest in this topic encouraged me to write this book.

This volume represents my current perception of how science is shaped by the vested interests it serves and by the society that supports it. I have tried to present a unified analysis of the influence of different interests on the day-to-day practice of research science as well as on the larger framework of scientific management and administration. But this book can also be seen as my own effort to escape the traditional model of scientific investigation – the reporting of a supposedly open-ended experiment as if pure curiosity were the only motive. Rather than presenting a paper, with Abstract, Introduction, Materials and Methods sections, Results Discussion, and Conclusions, I have conceived of this book as an illustrated argument of how a variety of vested interests affect scientific work. Like other investigators, I started out with a hypothesis and sought specific information to test it with. Most of my findings supported my suppositions; when they did not, I tried to understand why and to decide whether I should then doubt my initial assumptions. But, overwhelmingly, the research underlying this book led to the conclusion that science is in large part a function of the interests it can serve. I use a number of stories as illustrations, and I have attempted to rebut those who have disputed my analysis. But, inevitably, the text embodies many of my own preconceptions, values, and ambitions, for this is a personal statement, based on my own perceptions and experience. Readers should be aware of this perspective; I warn against accepting anyone's bias without critical analysis.

Presenting an argument for reform of the management and practice of scientific research, I have emphasized the weaknesses in the way that scientists usually work. Much of what I have to say is harsh and unpleasant. I have deliberately dragged scientists out of what remains of their ivory tower and subjected them to the kind of criticism normally reserved for businessmen and politicians. The chapters that follow are bound to irritate, and even offend many who devote their lives to the scientific enterprise, and I regret any pain that I may cause. It seemed to me worthwhile, nevertheless, to portray a tarnished image of modern science in the hope of stimulating public debate on these issues. Scientists are frail, and the science they practise is often flawed, misguided, or even corrupt. Without fundamental changes in the structure of scientific training, advancement, and rewards, we risk wasting one of our most valuable human and intellectual resources.

I do not, however, mean to cast moral or intellectual aspersions on the scientific endeavour, or on the scientific community. Very few, if any, scientists are deliberately dishonest, or stupidly ignorant of their political or corporate roles. Rather, they merely inform their work with the same assumptions, aspirations, and attitudes that underlie the rest of their lives. Because of the dedication and even sacrifice often demanded by scientific work, especially early in an academic career, there is likely to be an exceptional devotion to the goals of research in general and the problem chosen for study in particular. Such intense devotion may make it even more likely that scientists' research work – the way it is done as well as what it uncovers – is, in large part, an extension of their beliefs about what kind of information is important and how it should be described.

This book is intended for many different kinds of readers, from active scientists, science-policy advisers, and administrators to science teachers, students, advocates, journalists, and the general public. Recognizing the needs of this broad audience, I avoided producing an academic, historical, or philosophical treatise. Many authors could, and indeed have, restricted themselves to these approaches. In particular, Robert Merton, Harriet Zuckerman, Jerome Ravetz, John Ziman, Stephen Toulmin, Michael Polanyi, Alvin Weinberg, Richard Westfall, Paul Feyerabend, and Thomas Kuhn have all written extensively on these subjects, primarily for a scholarly audience. But my book is not for or about academic theorists. It is about the practical effects of the current administration and practice of science on political, corporate, and academic decisions, and on how these structures and procedures can be reformed.

For most people, the mystique of science is very much intact. Their primary exposure to it is through school courses presenting past discoveries, and media reports celebrating scientific advances and technical achievements. To this vast majority, it is important to point out that scientific endeavour is fraught with pitfalls and uncertainties, and that expectations and assessments of scientific investigations should be tempered accordingly. While it is true that individuals who are involved with particular technical issues that concern them personally (such as local environmental contamination) are becoming more sceptical of scientific reassurances, and that the recent technological disasters at Chernobyl and Bhopal and the space-shuttle explosion have also shaken public faith in technological solutions, scientific support for political, corporate, and financial decisions is still hugely influential.

This project was generously supported by the Canada Council's Explorations Fund, which enabled me to devote much of the past few years to researching and writing this book. (My two children, born during this period, prevented me from becoming too obsessed with the project!) During this time, Ursula Franklin has been a source of staunch encouragement and unusual vision. Numerous discussions and correspondence with my colleagues, students, friends, and relations have also challenged and changed my ideas as well as the focus of my work. Their backgrounds and contributions reflect the range of readers for whom this book is intended. In particular, the examples, recommendations, and emphasis I use and develop owe a great deal to the following people, among many others: Grahame Beakhurst, John Browne, Stephen Ceci, Jill Eisen, Bruce Gellerman, Michael Gold, Trevor Hancock, Stuart Hill, Brenda Kahan, Sally Lerner, Les Levidow, Bill Lewis, Brian Martin, Jerome Nriagu, Beverley Paigen, David Pecaut, Chris Roddick, Paul Ross, David Savan, Ted Schrecker, Ron Thorpe, David Wimhurst, Anne Wordsworth, Christopher Wren, Robert Young, and Stephen Young. Kathy Cooper, Tony Easty, Bob Gibson, Stevan Harnad, Evelyne Hertel, Walter Nelson-Rees, James Pirkle, Richard Royall, and Barbara Wallace were generous enough to read parts of the draft manuscript, making helpful suggestions or pointing out flaws or weaknesses in the text. Daryl Chubin, Art Forer, Jenny Fraser, Danny Henry, Trevor

Levere, and John B. Robinson each carried out a thorough review of an entire draft of this book, improving it immeasurably. Rose Sheinen gave a great deal of her scarce time to a meticulous and incisive critique of the manuscript, which prompted major revisions to the text. With insight and humour, Laura Roebuck showed me how to clarify my ideas and simplify my style. Much-needed moral and practical support was given by Valerie Hussey. Peter Livingston and his associate David Johnson, acting as my agents, were conscientious and efficient throughout. Being edited by Charis Wahl was a pleasure; she was always generous, encouraging, funny, and constructive. Lynda Gladysz and Darin Jacques carried out much of the bibliographic research, and Kate Hamilton and Ann Phelps were always tolerant, supportive, and amazingly accurate in typing and otherwise improving the text. In spite of all this assistance, I have no doubt that many flaws remain: the responsibility for all of them is mine alone.

Beth Savan
Toronto, Canada
February 1988

Chapter One
SCIENCE UNDER PRESSURE

Extensive scientific testing has shown that proper use of food irradiation does not present a health hazard. All of the evidence indicates that consumers have nothing to fear from irradiated foods but instead can look forward to a greater variety of high-quality food products if this process comes into more widespread use in the United States.

-- Kathleen A. Meister, "Irradiated Foods", American Council on Science and Health, 1985

My contention is and has been that a dose of radiation sufficient to cause changes in food pathogens to result in their death must cause comparable changes in the food itself.... The actual dose of radiation is irrelevant since a single carcinogenic insult is all that is needed to produce a malignant neoplasm a decade or more later. Thus, irradiated food should be safe only for individuals with a life expectancy of a decade or less....

-- George L. Tritsch, cancer researcher, Roswell Park Memorial Institute, letter to *Chemical and Engineering News*, July 21, 1986

We're all getting used to hearing two scientists give opposing answers to the same question, but it's still a very uncomfortable experience. It confuses and disturbs us to see two experts, examining the same problems with the same data, come up with contrary conclusions. Not only does such a situation make it much more difficult to resolve the important controversies the scientists are commenting on, but the scientists' contradictions also challenge the popular and rather comforting image of the scientist as selfless truth-seeker. They suggest that scientific information is equivocal – that its meaning may, like beauty, lie in the eye of the beholder.

If so, scientists are not simply discovering external facts but rather are, in a sense, manufacturing knowledge from their personal templates. The individual choices scientists make so limit their conclusions that two investigators can set out on the same research trail and arrive at entirely different destinations. If, meanwhile, we are seeking clarity and certitude in scientific results, we cannot help but be concerned when we find that the same basic research problem or set of data has spawned more than one answer to the question that matters to us.

Science and scientists have traditionally been held in high esteem, even revered. Earlier in this century, the public had unbounded faith in the power of science to explain mysteries and solve practical problems. As Richard Gregory, former editor of the journal *Nature*, put it: "My grandfather preached the gospel of Christ, my father preached the gospel of socialism, I preach the gospel of science."[1]

Indeed, the image of science and technology as a panacea for modern ills still holds a powerful appeal. We rely on scientific knowledge to cure our diseases, to help us grow crops more productively, to come up with sources of cheap energy. Our unprecedented understanding of the natural world, and our development of technology to control it, have enabled us to achieve incredible feats: individuals whose lungs are damaged can have them replaced and live for many years; through genetic engineering, we can now create new forms of organisms that never existed before; words, pictures, sounds, and other information can be transmitted from place to place and person to person instantly, and amazingly cheaply. In fact, current scientific knowledge is so vast that you virtually have to be a specialist to grasp even the rudiments of a particular field.

The power of scientific knowledge confers a special status on scientists. They are often portrayed as objective experts, purveyors of specialized information that they can use to make sophisticated technical judgements that we cannot or need not understand. When this happens, science becomes remote, incomprehensible, alienating, and even mystical for the lay public. The scientists, masters of the subtleties and complexities of their disciplines, can be seen as a secular priesthood delivering the "gospel of science", as Gregory calls it, to the faithful. It is particularly tempting to lean on expert authority in an era of extreme uncertainty, in which risks of global disasters are better known and more widely feared than ever before. How else can we tell whether nuclear-power plants are safe, whether we should treat a medical problem with drugs, and whether a new pesticide causes mutations or cancer?

The trouble with delegating this special authority to scientists is that, however convenient the image, they are not human encyclopaedias or data processors who need only find the right file to come up with the single true answer to our question. Like traditional priests, they are individuals with personal histories, childhood experiences, phobias, religious and political convictions, hopes, goals, desires, and ambitions. They are, like all of us, emotional beings, pursuing, usually with passion and even obsession, work to which they are intensely dedicated. Inevitably these human qualities colour a scientist's work and influence the field, line of research, methods, and ultimately the conclusions of his or her investigations.

Scientific investigation is not a straightforward act of observation and recording, but rather a complex series of personal choices and subjective interpretations. It is striking how divergent results can emerge when separate teams of scientists ask the same question, using different approaches. Depending on the assumptions and techniques employed to investigate the problem, they can produce dramatically different sets of experimental data. At other times, the meaning of universally accepted data may be a matter of dispute. In this situation, alternate methods of analysis and interpretation can draw investigators to contrary conclusions.

The choice of methods, assumptions, and interpretations is often not an idle academic matter, but one fraught with passionate controversy. In most fields, the academic, political, and philosophical inclinations of the scientist will strongly influence these choices. Together with the practical constraints on the investigation, such as the need for equipment and staff to carry it out, matters such as the allegiance of the investigators to the academic or political status quo will make a big difference to how they carry out their work. Do they hope to extend an accepted theory or to challenge it? Are they comforted or disturbed by the way their discipline is applied to current social controversies?

In the field of radiation health, for example, predictions of the risk posed by low-level radiation are based on extrapolations from much higher levels of exposure. Should scientists assume that radiation is proportionately as harmful at low doses as at high doses, so that a half-dose will give half the risk of cancer? Or should they assume that because cells may be able to repair radiation damage when it is slight enough, that lower doses are proportionately *less* risky than high ones? The information available on this matter is far from conclusive, so the way the data are interpreted, or the assumptions made about the behaviour of irradiated cells, becomes very important. In effect, such choices determine the conclusions drawn about the risks of exposure to the low levels of radiation released during nuclear-power production. More importantly, these assumptions and interpretations are vehicles for the scientist's own beliefs and values – for example, convictions about the need for electric power, for a high technological standard of living and about the society and way of life that go along with it.

Ursula Franklin, a Canadian metallurgist and a prominent commentator on the role of science in society, calls the intellectual, emotional, and social baggage that scientists bring to their work "the context for science".[2] In this book I investigate exactly how this personal context can influence the outcome of scientific research. One by one, I identify and examine the various beneficiaries of such bias: scientists themselves, the government departments or corporate bodies they work for, and the established science administrators and laboratory directors who can be rewarded by the shrewd use of science to favour certain ideas, theories, or fields.

In the chapters that follow I attempt to illustrate and analyse how this process works. In some ways this book recapitulates the scientific-research process that is its subject. My working hypothesis sprang from a personal experience that caused me to question the motives behind ostensibly objective scientific judgements. I began to notice how scientists' ambition, convictions, and preconceptions about what kind of results would be valid or useful coloured their thinking and their work. And, as I gained experience in the field of environmental science, I learned how scientific data often became the arbiter of social or political issues. My interest in the politics of science led me to explore the information on these subjects avidly, and, based on my findings, I developed the model of how scientific research can serve academic and professional ambitions that is presented in Chapter Five. The data that I collected suggested a kind of purposive science, embedded in a very political research environment, where the research work undertaken can serve a host of internal and external vested interests. Throughout this process I have sought information that suggested contrary theories, and have repeatedly invited critical analysis of my work, through a CBC radio series, through several lectures and articles, and through a host of conversations, series of correspondence, and reviews of draft sections of this manuscript.

The evidence that I was able to collect, some of which is cited in the endnotes, supported my initial scepticism about the motives of scientific researchers. Indeed, it indicated that these people are subject to numerous pressures, which constrain and can even direct their work. It led me, ultimately, to re-evaluate my own image of scientists and the tasks they perform.

If we admit that scientists often produce knowledge that is imprinted with their own ideas and values, then we have to change our expectations both of scientists and of the work they do. We must not confine our interests to the hard data and their implications for our own lives and for public policy. We have also to explore the history of the information: who did the work, why, and for whom. Perhaps most important is the question of who stands to gain by particular findings.

This book, then, is unashamedly partisan: it embodies an argument for exposing the vested interests that drive scientific research, and for reforming scientific administration to encourage more egalitarian, diverse, and constructive research. Not only does the practice of science by scientists have to change, but its relation to the public must also be transformed. Just as a patient has a right to informed consent to a medical procedure, we, the public, should demand the right to informed consent or refusal of new chemicals, technologies, or treatments.

Recognizing the importance of the disciplines underlying many current controversies, the media have begun to pay more attention to scientific research, interviewing investigators and engaging regular science commentators. Between 1977 and 1984 fifteen magazines, seventeen television shows, and eighteen newspaper sections covering science were initiated in the United States alone.[3] Increasingly, the news depends on science. Decisions on all kinds of issues – from how to control cockroaches in public housing, through how to treat mine tailings, to how best to regulate soft-drink containers for the public good – are made on the basis of expert scientific advice. Indeed, science has become an integral part of modern politics. Accordingly, corporate executives, politicians, and lobby groups alike justify their positions with scientific results. And as scientists come under increasing pressure from these parties, the research work they undertake is inevitably altered.

Depending on the political and academic climate, certain research topics will become more attractive because of the prestige and funding that flow to investigators in those areas. Moreover, the assumptions, method, and analysis of the work will, unavoidably, be coloured by the personal and professional rewards to the scientists, which are contingent upon the outcome of the research. The informed public is recognizing some of these pressures, and people are becoming somewhat sceptical of science. (In 1980, the U.S. National Science Board found that only two-thirds of the general public thought that the benefits of scientific research outweighed the harmful results, compared to 87 per cent in 1957.[4])

The community of scientists is also becoming more self-critical. Recently many professional scientific bodies have developed codes of practice to guide researchers and administrators. Others have developed defences of scientific objectivity and integrity. In part, this is a response to the scientific frauds that have been greatly sensationalized over the past few years. Numerous newspaper and magazine articles[5] and a popular book, *Betrayers of the Truth*,[6] have pointed out that scientists can be ambitious and opportunistic, and sometimes use unscrupulous methods to achieve academic rewards. Corporate science has also received a public battering in a number of books investigating the suppression of evidence by the major asbestos manufacturers of the occupational hazards of working with asbestos,[7] and the corporate retribution meted out to scientists and others who expose product flaws.[8]

All of this evidence for the human fallibility of the scientific endeavour merely shows that scientists are no different from other professionals. Because we aren't used to thinking about scientists this way it may shock us, but we shouldn't infer that scientists are dishonourable. If anything, scientists may be more dedicated, honest, and open than people in fields where the rewards and the overt financial or political pressures are greater. The ethic of selfless, open, and single-minded search for the truth, though largely mythical, is still taught to young scientists;[9] and it appeals to and creates a sense of obligation for the profession as a whole.

Cases of corruption are still rare, and incidents of scientific fraud may be increasing only because there are now more scientists than ever before, and more journalists eager to report scientific misdemeanours. Whether the *proportion* of scientists engaged in deceptive research practices is any higher now, we cannot know. What these occasional examples of invented, altered, or plagiarized data do show is how researchers can cheat, knowingly, to secure a successful career.

What has not been so well publicized is the more pervasive but less obvious manner in which scientists can deliberately or, more often, unconsciously work in such a way that their conclusions are bound to support a particular position, policy, or action.

Philosophers, sociologists, and historians of science have argued that social, political, and personal factors heavily influence the prevailing dogmas accepted by the majority of the scientific community as well as the interpretations of experimental work carried out by individual researchers. Thomas S. Kuhn, a historian of science, shocked the academic community twenty-five years ago with his theories on what he called "scientific revolutions", during which one explanation or "paradigm" for how nature works is replaced by a different but not necessarily truer one.[10] During the periods between these intellectual upheavals, he argues, novelty and challenge to the prevailing dogma are suppressed and resisted. Paul Feyerabend, a philosopher of science, takes this idea farther when he presents scientific investigations as a largely irrational process leading to competing theories, none of which has a corner on the truth.[11]

At the level of the individual investigator, Richard Westfall, a historian of science at Indiana University, has demonstrated how Isaac Newton, discoverer of such fundamental physical principles as gravity, the spectral composition of light, and the mechanical laws of motion, exaggerated his achievements by using what Westfall calls a mathematical fudge factor![12] Newton consistently corrected his raw data to remove inaccuracies or vagaries and to make it appear that his calculations were uniformly exact and precise. Newton's manipulations were probably quite conscious, but Stephen Jay Gould, a professor of geology at Harvard University and author of several collections of essays on the history of science, has described a vivid example of unconscious data-fudging: Samuel Morton, a physician working in the mid-1800s, took the prevailing attitudes regarding white people's intellectual superiority so much for granted that he published obviously biased analyses of his raw data to support this theory. He believed that the various human races were really separate and unequal species, and by a series of inconsistencies, errors, omissions, and miscalculations he managed to interpret his data as supporting this opinion.[13] Yet, ironically, when Dr. Gould revealed Morton's errors he inadvertently misread one of Morton's figures, which led him to undervalue the average Caucasian skull size and thereby reduce racial differences in accordance with his own preconceptions.[14]

In recent years there has been a growth of interest in the social role of science and the behaviour of researchers. Other students of all these topics have reviewed and analysed aspects of science management and academic fraud, corruption in industrial science, and self-deception in past, or, occasionally, current contexts. Scientific fraud has been attributed to academic ambition, given too free a rein by the system of career advancement;[15] or to exceptional dishonesty in an unusually upright profession.[16] Corporate corruption of science is often explained by the "scientists are human" homily, and likened to corruption in other fields; stories of self-deception have been treated either as cautionary tales of investigators diverted from seeing the truth by fairly obvious errors[17] or, on occasion, as consequences of the political and economic roles of science in modern society.[18]

In this book I make an effort to deal with all these issues in a unified manner. I develop an analysis of these phenomena and identify the scientific choices made during research, revealing the way that personal, political, and financial vested interests sway these decisions. I show how the self-regulatory mechanisms established by the scientific community fail to deal with the biased conclusions that often result from the research process.

No attempt has been made to survey all the incidents of scientific malpractice, or to review the rich academic literature in this field. The human activity called science has been studied with increasing interest over the last twenty years, using a variety of theoretical approaches. The psychology, sociology, philosophy, and history of science are indeed recognized subdisciplines. While several of the most fascinating and insightful studies originating in these fields are cited above and in the pages that follow,[19] this book does not attempt a thorough survey of the work in these fields.

Also, the philosophical debate regarding the validity of scientific endeavour has, as much as possible, been avoided. The question of whether there is any one true pattern of reality is of less interest to me than what results, socially and politically, from a particular scientific perspective, and I have tried, throughout, to emphasize this pragmatic approach. For the purposes of this book at least I accept the scientists' view that they are "discovering" external truths: I do not, therefore, dispute that $E = mc^2$. Instead I would rather ask whether the effects of a new pesticide should be measured in terms of reduced crop loss or contamination of agricultural runoff, and whether the possibility of wrongly concluding that a new pesticide causes no human health effects should be given the same statistical weight as wrongly concluding that it does. In other words, I have made an effort to bring a practical perspective to the problem of how scientists and the data they generate are, in part at least, products of internal and external pressures. Each of the following chapters focuses on a detailed description of one particular story, with other examples or statistics raised as evidence of the problem's prevalence or as an aid to analysing its roots.

Chapter Two demonstrates how science can be an elusive, plastic force, clarifying puzzles for some scientists while obscuring them for others. The same experiment performed by different people can yield totally dissimilar results, because of contamination, error, or impurities. In other words, we cannot take it for granted that science is uniform, that scientists are consistent with one another, or even that the materials they use are what they seem. Discrepancies, although usually inadvertent, and deeply regretted once the mistakes are discovered, do happen frequently.[20] This chapter shows how even the most impartial and conscientious researcher may use impure samples, cell lines or animal strains, which result in misleading conclusions.

The choice of assumptions and methodology used to answer a research question can also profoundly influence the results obtained, and these decisions are clearly dependent on personal preferences as well as scientific rigour. As a result, debates regarding experimental assumptions can have far-reaching social and even political consequences. Chapter Three follows two examples in which technical and academic disputes paralleled public controversies concerning environmental pollution and public health. This chapter explores the borderline between unconscious self-deception and more deliberate steering of research results. When scientific investigators become embroiled in debates, there can be a strong inclination to find the results that are expected or desired. The two cases analysed in Chapter Three illustrate how the way a question is posed, and the choice of research materials, methods, assumptions, and interpretations can lead investigators to the results and conclusions they find easiest to accept.

Chapter Four reveals the obvious benefits that corporations can derive if academic institutions devote more of their time to lines of investigation that might result in corporate profits. This "bottom-line" approach has prompted heavy investment in university biology departments across the United States by drug-manufacturing and genetic-engineering firms. On a smaller scale, scientists have a strong vested interest in channelling their patentable findings into their own companies for development and sale. Chapter Four examines obvious cases of scientific corruption that occur when direct financial incentives are involved. The coexistence of scientific-development companies and academic activities in the same laboratories is becoming a strong but not always welcome influence on the training, communication, and direction of modern researchers. Even more disturbing is corporations' investing in entire university departments: they are quite possibly skewing the direction of academic research in many fields as well as more general science-policy and regulatory decisions across the North American continent for years to come.

Researchers can also further their careers by supporting, deliberately or otherwise, the hypotheses, theories, or fields they study. Taken to its extreme, they can even fake or deliberately stretch results to make them more impressive. In today's highly competitive scientific world, jobs are scarce and honours accrue to the very few workers who not only are prolific, but contribute work perceived to be original, important, or critical for investigators outside of a narrow specialty. Under these circumstances the temptation to hoist oneself up the ladder by using improper research techniques is great. In Chapter Five, an examination of the power of publications, the unbalanced structure of research laboratories, and the influence of the Old Boy network is integrated into an analysis of why research irregularities seem to be a fact of modern scientific life. The same system of scientific advancement and rewards that ensures that certain ideas, fields, and individuals prosper at the expense of others engenders not only the latent self-deception to which most scientists fall victim, but also more serious and rarer cases of out-and-out corruption.

The management of academic science is further explored in chapters Six and Seven. The system of research funding and publication relies on the participation of a relatively small circle of established researchers. Naturally, they promote themselves, their friends, and the ideas, theories, fields, and problems that they prefer. These chapters show how young scientists are encouraged to work in large successful laboratories on popular problems and to share some of the credit they may get for their findings with their laboratory directors, who are already established. The influence of certain people, ideas, and vested interests in the field of science as a whole is thus perpetuated, making the role of maverick a daunting one in an enterprise that relies for its ultimate advancement on the challenge of new ideas.

Such use or abuse of science can divert a whole field, and it may be years before the weight of evidence indicates that the work ought to be reassessed. Meanwhile, a great deal of time and money may have been wasted in following false scientific leads. But much graver consequences are also possible: dangerous drugs may be prescribed, hazardous pesticides applied, or an endangered species harvested to extinction as a result of faulty data. The social consequences of scientific corruption make its unmasking and reformation matters of great urgency.

It is hoped that, whatever the shortcomings of this book, it will stimulate further discussion of scientific-research practices and of how best to reform them. It would be naive and even misguided to wish that the personal, political, and corporate influences on science did not exist – scientists can't and shouldn't isolate themselves in a secure ivory tower in the midst of a very insecure society. The final chapter of this book analyses the similar ways these different vested interests operate, and suggests some basic safeguards for the scientific enterprise. Scientists challenging the status quo should be protected; innovators exploring new ideas or new fields should be encouraged; and a diversity of perspectives on scientific issues of importance to the public must be funded. We have to develop guidelines for the wise use of public money and for more balanced investigation of controversial political issues. Finally, we can foster a sceptical attitude on the part of the public, the politicians, and the scientists themselves, by openly admitting the personal nature of scientific investigation and its reliance on the values, interests, and preconceptions of scientists and the society in which they live. Only when we acknowledge this often unavoidable bias can we modify our expectations of what scientists can find out for us, and how we can best use their findings.

Chapter Two
SCIENCE AT RISK

The non-scientist may assume that scientific research is the relatively straightforward process of asking a specific question, designing an experiment to answer it, and then passively receiving results that provide a clear response to the research question. But the practice of scientific research is never this simple. The trick – some would say the art – of science is to make sure that the experimental results answer the question posed rather than an irrelevant or different one.

To find out about the general, often theoretical issues at the heart of most scientific research, investigators develop research hypotheses relating their specific experiments to the broader problem that interests them. To achieve unequivocal, relevant results, the hypothesis should be framed in such a way that different experimental outcomes lead to different conclusions regarding the larger issues. Otherwise the experiment will be useless, producing ambiguous or misleading results. (This would be the case if, for example, the hypothesis that a particular chemical is not carcinogenic were tested by exposing mice to that chemical for such a brief period that no ill effects on those animals would be manifest.)

Furthermore, hypotheses can only relate specific experimental results to more general situations by assuming a number of similarities between the experiment and other relevant work in the field, and between the experiment and the "real world". Any weakness in these assumptions, which relate to the purity of materials and the consistency of methods used or the behaviour of the experimental subjects, can invalidate or seriously limit the conclusions drawn from the work. And problems of purity are virtually unavoidable in a research method that requires total artificiality – the isolated measurement of the effects of controlled variables.

For example, if we want to find out what effect a certain chemical has on human health and, in particular, whether it is likely to cause cancer, we might expose a battery of mice to this chemical. Obviously, all kinds of things affect the disease rates of wild mice – their food, their genetic makeup, their habitat, and the weather are only a few of the factors that would profoundly influence their health. Yet, experimental success often depends on the elimination of natural influences – the reduction of reality to an artificial test of the effect of one pure and independent variable on a known system. Therefore, in order to get meaningful results, we would have to know all about the mice being tested, and the food they were fed, to be sure that the mice weren't especially susceptible or resistant to cancer and that the food had no effect on their chance of contracting cancer. We would have to know the exact amount of chemical being administered, and that it was pure and did not include traces of any other chemical that does or may cause cancer. We would have to know how long to wait before checking for the appearance of disease, to be sure that symptoms had time to develop, and so on. In brief, we would have to control and standardize all these factors to be sure that the results were reliable and could be compared with the results of similar tests, using other chemicals. (The goal of experimental standardization is, of course, only achieved at the expense of relevance to the real world; the purer, more artificial the laboratory setup, the less certain we can be that the results have any meaning for complex, inconsistent nature.)

In field investigations, the analogous focus on one narrow question requires a similar discipline. Very clear schedules for observations, consistent notation, and sophisticated analysis are necessary to take into account the influence of as many factors as possible. For example, the season, time of day, weather, and other variable effects must be controlled carefully, or, ideally, kept constant. (Yet the real world tends to intrude, despite all the precautions that the scientist might take!)

Needless to say, this task is enormous, if not impossible, and the clear-cut evidence that scientists long for may prove elusive. Apparently reliable findings may be fundamentally flawed as a result of contaminated materials, incorrectly calibrated instruments, or, worst of all, an experimental design that is biased towards expected results. Such slips, oversights, and false assumptions can easily skew scientific results, from which researchers will draw entirely misleading inferences, often despite their best efforts to carry out accurate and objective research.

Many of the decisions regarding experimental materials, methods, and assumptions are dictated by the traditions and precedents in the field. More significantly, however, these choices can allow scientists to steer their findings in a predetermined direction, either deliberately or unwittingly. This chapter demonstrates the ease with which inadvertent errors or false assumptions regarding the quality of experimental materials are made, with consequences both political and scientific.

In the opinion of some experts, inbred, genetically pure mice are the single most important tool in the search for a cure for cancer. One geneticist, Hans Grüneberg, goes so far as to assert that "the introduction of inbred strains into Biology is probably comparable in importance with that of the analytical balance [a very precise weigh scale] into Chemistry."[1] These mice, which are the offspring of brothers mated with sisters for twenty or more generations, have virtually identical genes for certain traits. As a result, these inbred mouse strains are in great demand by researchers who want to study the reaction of this species to various treatments, without having to take into consideration differences between individual mice. Like a brand-name drug or chemical product, inbred mouse strains, such as the BALB/c stock, should be homogeneous, and buyers should be confident that each individual BALB/c mouse is genetically identical to every other one.

In an article published in 1982 in the widely read and reputable journal Science, Brenda Kahan, a zoologist at the University of Wisconsin, and her colleagues alarmed biomedical researchers across the continent by announcing that the mice she was sent were far from the uniform BALB/c stock they were supposed to be.[2] On June 29, 1983, she filed a lawsuit against Charles River Breeding Labs Incorporated, the largest supplier of scientific research animals in the world. Charles River's publicity material boasts that its "commitment to leadership in the field of quality assurance has helped establish [it] as the world's foremost breeder, and earned [it] the proud distinction of 'The International Standard' ® for laboratory animal excellence".[3] (In 1982, the company's net sales approached $41 million.) As a result of the genetically impure BALB/c stock supplied by Charles River, Kahan claimed that nearly a year and a half of her precious research time had been wasted and important research invalidated. She sued the company for more than US$200,000.

Kahan discovered the problem with the mice because her work happened to involve testing for the presence or absence of a particular gene. Kahan was attempting to produce "chimeras", that is, mice that develop from cells originating from different mouse strains: the resulting animal is made up of cells that could come from either strain. Chimeras represent a powerful research tool, which, in different contexts, can be used to investigate embryonic development, the behaviour of cancer cells, and the behaviour and treatment of animals with specific genetic traits or deficiencies. In Kahan's case, by following the development of BALB/c mouse embryos injected with tumour cells from a different mouse strain, she could learn about how the tumour cells operated. Kahan identified the genes for each mouse strain by picking out several characteristics, both biochemical blood enzymes and visible coat-colour traits, which differed. When she tested the animals' blood-enzyme types to find out whether the newborn mice did indeed display characteristics of both embryo types, she was mystified to discover that their qualities differed from the usual characteristics of *either* donor strain. She checked the donor BALB/c mice she had been sent from Charles River Labs and, to her dismay, she found that they weren't typical BALB/c mice at all. To test her suspicions, she ordered another batch of BALB/c mice and went through her battery of blood electrophoresis tests on the new animals. To be absolutely sure, several animals were forwarded to colleagues at the University of Minnesota, where the tests were repeated. All this work, which took several months, supported her worst fears: the mice were definitely not inbred, genetically pure BALB/c stock.

Kahan's lawsuit, one of the first of its kind, challenged the biological supply house to accept responsibility for this mishap and warned other suppliers that failure to guarantee their products could result in serious financial and public-relations consequences.

Her legal complaint charged: "At all times relevant to this action, Charles River knew or should have known, from its own testing or from others' reports, of the genetic defects in its mice. Charles River recklessly failed to notify Kahan or anyone at the University of Wisconsin-Madison about the defects in its mice."[4]

The implications of this lawsuit are significant. Genetically pure mice, and the BALB/c stock especially, are being used increasingly in the biological-research community. National Public Radio reported that "[Charles River Labs] company officials once estimated that as many as 112,000 suspect BALB/c mice have been sold."[5] They announced that as many as a thousand researchers might be affected, and that they had talked to more than twenty scientists whose research had been dependent on using pure BALB/c mice. Of course, not all of these researchers would necessarily have had their research invalidated by using genetically impure mice. But for some, like Kahan, the effect was devastating: "Our first thought was,

'Maybe our experiments succeeded wildly,' but that was unlikely ... so quickly we went from 'This is fantastic' to 'Well, this is too fantastic, there must be something wrong,' and so when the resolution of our experiments improved, sure enough, it was obvious that there was something seriously wrong with the mice."[6]

Kahan alleged that the failure of Charles River Laboratories "to notify [her] or anyone at the University of Wisconsin-Madison about the defects in its mice"[7] was reckless. Charles River responded that she should, in turn, have notified them when she suspected contamination.[8] On January 20 Kahan submitted an article to *Science*, a very widely read journal. The usual peer-review practices delayed publication by several months, and the research community learned of her warnings to other scientists using BALB/c mice in the July 23 issue of *Science*. (For more on peer review and publication problems, see Chapter Six.)

The article concluded with this warning: "The seriousness of our findings cannot be overemphasized. Since shipments received in January 1981 and September 1981 from the Stoneridge facility and in September 1981 and October 1981 from the Portage facility were incorrectly identified it may well be that shipments in general made from these facilities over many months may have led to erroneous conclusions in research experiments... results of experiments on NK activity, tumor susceptibility, and immune responsiveness may need to be reassessed."[9]

Kahan was fortunate that her work required the genetic identification of the animals she used. Many researchers who depend on BALB/c mice don't do this kind of check. As Dr. Clifford Ottaway, a medical researcher at the University of Toronto who uses these mice, pointed out: he "wasn't aware of problems with the mice" from his own studies, as it wasn't "part of his work to formally check the genetic status". His work didn't require uniform stock, and so no conclusions from his experiments would have been affected if the mice hadn't been BALB/c.[10] But this does raise the very real possibility that other scientists, whose research is less dependent on genetic purity, may not learn about or publicize contamination problems. While their research wouldn't necessarily be invalidated by the use of genetically impure mice, genetically varied animals do introduce another uncontrolled factor into experiments. If unusual or unexpected results were produced, the genetic factor would have to be discounted before more interesting explanations could be explored. Certainly the extra standardization in experimental method that inbred mice provide must have some value for these researchers, or they wouldn't have been willing to pay for them roughly double the price of outbred mice.[11]

The surprising fact is that only a handful of the hundreds of researchers who used the impure BALB/c mice have admitted doing so. As most of the others probably never checked the identity of the mice they used, we have no idea how many research results are suspect and what published data should now be questioned. The list of researchers who received Charles River BALB/c mice from the same batches as those sent to Kahan has not been made public. Several scientists first heard of the alleged contamination through newspaper reports,[12] and it may be that many researchers are still unaware of the doubtful purity of their stocks. A more cynical conjecture would be that the scientists are aware of these problems but haven't been willing to admit that their work should be reassessed in the light of this information.

The Kahan case is a striking one, but it's certainly not unique. Dr. Ottaway explains that although this sort of genetic contamination is "unlikely in the normal course of events ... [since] the history of animal breeding houses has been to devise strategies to prevent this ... accidents can always happen, potentially very easily."[13] In facilities crammed with different strains of mice, it requires just one animal to escape into another breeding area for generations of mice to be contaminated.

It happened before at Charles River. In 1980, Dr. Heinrich Bitter-Suermann reported that the rats he received for his experiments on skin and pancreas transplantation were not genetically pure.[14] These rats had been supplied to him every month at the McGill University Cancer Research Unit in Montreal, Quebec, and then at Georgetown University in Washington, D.C., by the Charles River Breeding Labs. The experimental results were so variable that he finally attempted to graft spleen and skin tissue from donor rats onto supposedly identical rats supplied on different dates. To his astonishment, many of the grafts were rejected; this led him to check the genetic identities of the rats. Sure enough, tests revealed that they were not absolutely pure inbreds.

Despite this precedent, Kahan doesn't believe that Charles River was *necessarily* malicious in distributing impure mice. Her charges laid out three alternatives: that Charles River Labs didn't test to ensure that the mice were pure; that they tested in an improper fashion and didn't discover the impurity; or that they tested, learned the mice were impure, and knowingly distributed them under false pretenses.[15]

Charles River wasn't necessarily deceitful, but may have been negligent. Although a finding of deliberate fraud might result in a higher fine than unconscious negligence would, the critical judgement to be made concerns responsibility.

In August 1984, Dr. Alvin Warfel, a cancer researcher formerly at New York's Sloan-Kettering Cancer Center, brought another suit against Charles River for supplying impure mice.[16] He was quoted in *Science* as saying: "It's incumbent on the company to sell the product they're advertising.... In a certain sense, to me there's no difference between sodium chloride and BALB/c mice. That's why these companies exist, because they supposedly sell things of a certain quality. If researchers had to check all the reagents they purchased, they'd never get any research done. If they can't guarantee they're selling BALB/c mice when they say they are, why are they in business?"[17]

Unfortunately for other possibly affected researchers who would have been well served by a clear judicial finding on this matter, both Kahan's and Warfel's cases were settled out of court.[18] Meanwhile, Charles River has set up its own monitoring laboratory and has notified clients of at least one subsequent case of contamination.[19]

An editorial in *Nature* warned of a different, but equally serious problem:

> Those who work with cells in culture are continually aware of the risks of contamination. Especially in laboratories specializing in this work, one established cell culture may be contaminated with cells from another by a variety of means – contaminated glassware, people's hands or defects of laboratory plumbing. Almost by definition, the most esoteric kinds of cells, requiring the most elaborate mixture of nutrients to make them grow, are those which are the most susceptible to contamination.... Naturally, people with cells at risk are constantly on the look-out for contamination. This is obviously easiest when the cells at risk can be easily recognized, perhaps because of some distinctive biochemical characteristic. Again, however, and almost by definition, novel types of cells are likely to be least easily recognizable. In other words, it is entirely possible that people may work for weeks or months with cells from a tissue culture without knowing that the originals have been supplanted.[20]

A notorious example of cell-culture contamination illustrates the dire international consequences that an inadvertent slip can cause. In this particular situation, the cards were stacked against the diligent researcher by a remarkably vigorous, persistent strain of cells. In 1951, Henrietta Lacks, a thirty-one-year-old Baltimore woman, contracted cervical cancer. The cancer cells resisted all treatment and spread rapidly, killing her within a year. Meanwhile, samples of these cancer cells were cultured for use in medical research. Their astonishing vigour enabled the cells, dubbed "HeLa", to grow continuously in test-tube culture, outside the human body. As they were among the first such successful cell-culture lines, they were in great demand by researchers around the world, who were eager to work on standardized cells.

But wherever they were sent, it seems, these HeLa cells defied control. Sloppy laboratory techniques allowed them to invade hundreds of other cell lines, where they dominated and eventually destroyed the original cultures within months or even weeks. The cell cultures, now containing only HeLa cells, had changed but the labels had not; researchers used the mislabelled batches of cells in ignorance of their true identities. Scientists in five medical centres in the Soviet Union, and in laboratories as distant as Australia, Switzerland, and China, were all using HeLa cells in the belief that they were other cell lines altogether. Results of work on these mistaken cell lines became irrelevant or even misleading, since the research gave no insight into the workings of cells from other animals or organs than the human cervical cancer cells.

Dr. Walter Nelson-Rees, a cytogeneticist, has made the unmasking of these mislabelled cell lines a personal mission. In a series of nearly twenty scientific articles[21] over the past fifteen years, he and his colleagues have pointed their sometimes unwelcome fingers at dozens of cultures maintained by many respected researchers. Claiming that his revelations represent only the tip of the iceberg, he explained in an article in *Science:* "We have cited only major instances of cell cross-contamination, but it must be emphasized that such events occur quite frequently. Undoubtedly many cases go unnoticed or are detected instantly, and corrected, in the course of experimentation.... The risk of contamination or overgrowth of cultures by unrelated cells is a potential and often recurring problem where cells are grown and studied."[22]

As a result of cell-culture contamination, experiments on allegedly human cells to find out about our physiology, biochemistry, or disease, have in reality examined mouse, rat, or monkey cells. This work becomes "one of a kind" research, with correspondingly limited relevance for human biology and often for science as a whole. Dr. Nelson-Rees is tackling this problem in an aggressive, forthright, and (some might charge) obsessive manner. Yet his mission is worthwhile. Individual cases of contamination are not always publicized, even when they involve published research; but Dr. Nelson-Rees is willing to blacklist particular cell cultures, in named laboratories, used in specific published work. While the ever-present hazard of cell-culture contamination is generally accepted, it is quite another thing to admit that it may invalidate your own published research.[23] Correct and public identification of current and past cell lines used would no doubt be greatly delayed if it were not for persistent, scrupulous scientists like Dr. Nelson-Rees.

Biologists such as Brenda Kahan and Walter Nelson-Rees are not the only scientists affected by false assumptions about the purity of experimental materials. If anything, physical scientists are even more vulnerable, especially those attempting the very precise, accurate measurement of tiny quantities of materials. The enormous technical difficulties involved in obtaining clean samples of polar ice and correctly asssessing the minuscule amount of lead in them provide a case in point.

The debate on levels of atmospheric lead has aroused passionately held beliefs concerning the hazards of current levels of lead in the air. If it can be established that historical lead levels were orders of magnitude lower than those of today, which include contributions from industrial and automobile exhaust, then one might deduce that human biology did not evolve to deal with this increase in lead contamination, and it may therefore be harmful. Conversely, if high levels of atmospheric lead have been present for centuries, the recent additional lead pollution is less worrisome. Take this logic one step further: evidence of low lead levels in the past would lend strong support to the lobby for mandatory reduction or even elimination of the lead content of gasoline and food tins. Less stringent regulatory measures can be defended, however, if lead levels seem to have been relatively high for many centuries.

This debate has been prominent in the highly credible, if rather specialized, geological journal *Geochimica et Cosmochimica Acta*. In an experiment designed to measure historical atmospheric lead levels, scientists from the Central Laboratory for Radiological Protection in Poland measured the tiny amounts of lead and other metals present at different depths in permanent ice and snow deposits by boring out sample cores from glacial ice. They conjectured that early atmospheric lead would have settled on the earth's surface and, eventually, would have been covered by later accumulations of ice or soil. The depth of the ice sample indicated the approximate geological date of dust deposition. In a paper presenting the results of their study, Z. Jaworowsky, M. Bysiek, and L. Kownacka assert that they "did not find evidence of changes in rate of metal deposition during the last three decades, as compared with pre-industrial period, [sic], however, our samples of pre-industrial ice might be contaminated in part by contemporary fallout migrating from the exposed surface of old parts of glaciers into the deeper ice layers."[24] They also conclude that natural background lead levels clearly overwhelm the lead emissions from human industrial activity; these, they say, are "of minor importance for the pollution of the atmosphere on a global scale".[25]

A fierce rebuttal by Clair C. Patterson charges that the lead present in the air when the samples were taken and analysed would have infiltrated the ice, inflating the lead readings and leading to artificially high estimates of historical atmospheric lead levels. He alleges that serious flaws in the method of sample collection and the analysis techniques permitted significant contamination of the glacial cores. As a result, he claims, the findings of Jaworowski et al. are "not trustworthy", and their conclusions are "erroneous".[26]

Feelings run high on both sides of this debate, and the issue of potential contamination may be clouded by each scientist's preference, whether personal or political, for a certain scientific outcome. Clearly, the resolution of this seemingly arcane debate over tiny amounts of lead in old ice could have important implications for modern-day environmental regulation. Conceivably, the awareness of these implications could have influenced the scientists, consciously or not.

The hotly contested debate on historical lead levels may well prove to be, like the widespread contamination of tissue cultures, a consequence of understandable, if inexcusable, sloppiness: the unconscious relaxation of strict scientific procedures. In a case like this, the distinction becomes blurred between the understandable faith that scientists have in the results they have obtained and their allegiance to hypotheses they support. When does the routine performance of lab procedures, in the usual way, become the more deliberate steering of research results, so that they support a given conclusion? (See Chapter Three.)

Clair Patterson, at least, is known as a fierce advocate of stringent controls on lead. He is committed to highlighting "the extent of lead pollution in the foods and bodies of Americans. Today most chemists, including some in Federal laboratories, would report lead concentrations 10 to 1000-fold higher than true values if they attempted to collect and analyze lead...."[27] Dr. Patterson may be distinguished by his strong views on industrial lead pollution, but his scientific reputation rests on his excellent work as an analytical geochemist. On the basis of dating techniques he managed to produce a remarkably reliable estimate of the age of the earth. His work on polar ice appears to have achieved similar credibility. The consensus of scientific experts is now that Dr. Patterson's data are valid: the higher lead levels on old polar ice were indeed a result of contamination.[28]

The examples of experimental contamination cited above – of inbred mice, cell cultures, and polar-ice samples – differ enormously in their importance, in the motives and consequences for the scientists involved, in the implications for the research community in general, and in the public-policy judgements that derive from them. Yet they all raise similar problems of scientific quality control. In all disciplines involving experimental work, these problems will exist. Their importance will vary with the degree of standardization desired, and the risk of contamination affecting the experimental factors. In any case, they cast an element of doubt onto the scientific conclusions drawn from these studies.

As investigators become increasingly dependent on highly sophisticated research tools, such as genetically pure animals and measuring devices that accurately report tiny quantities of substances, the consequences of experimental contamination become more serious. Whole fields of investigation, such as the highly specialized work of cancer research, are standardized internationally, in a variety of research laboratories, by the use of identical procedures, techniques, machinery, chemicals, cells, and animals. The U.S. National Academy of Science has emphasized, in its report on Laboratory Animal Management Genetics, the advantages of "performing experiments on animals genetically identical with those employed by other investigators in other laboratories."[29]

Only when the experiments reported by different researchers are conducted in exactly the same way can their results reliably be compared. And only when these investigations are entirely consistent can the field be advanced, by building one scientist's results on those produced by another in a different laboratory. It takes only one discrepancy, the impurity of a chemical, an inaccuracy in the amount used, a minor problem with the analytical techniques or machinery, or experimental animals that are not uniform to completely alter the meaning of an experiment or even a whole series of investigations. The research may then become useless because the data depend on an unusual mix of experimental conditions.

This extreme dependence on a degree of standardization that cannot comfortably be taken for granted raises crucial questions concerning responsibility, notification, and compensation. Herbert C. Morse, an expert on inbred mice at the National Institutes of Health, and editor of the book *Origins of Inbred Mice*, considers that suppliers must guarantee their products, be they mice or chemical reagents. After all, he says, "the scientists are buying a product because it has the company's stamp of approval, and it should be what they say it is."[30] Clearly, if individual researchers had to check on the identity of all their supplies, much time would be lost on this activity. The supplies would be virtually useless until their authenticity had been confirmed.

Several suggestions have been made to deal with this problem. Michael F. Festing, a scientist at the Medical Research Council Laboratory Animal Centre in the United Kingdom, investigated the authenticity of twenty mouse colonies from ten different breeders in England. After discerning three major cases of inaccurately labelled breeds, he recommended that an independent genetic-monitoring facility be established to inspect animal breeders and test animals on a regular basis. He observes that scientists "go to considerable lengths to ensure that their chemicals are pure and glassware clean. They should pay equal attention to the purity of the animals that they use."[31] He recommends that breeders of colonies found to be authentic should be given a "genetically monitored" status, which they could advertise to assure users of their quality control. Festing doesn't suggest how this scheme could be funded, but a levy on breeders could support the tests. To ensure the independence of the testing facility, the levy would have to be paid as part of a long-term contract, possibly to an intermediary, such as a government agency, so that negative test results wouldn't threaten the financial stability of the testing laboratory.

The responsibility for quality control is more difficult to assign with cell-culture laboratories, which are smaller than animal-breeding facilities and are more widely dispersed throughout the research community. Ideally, those in charge of cell-culture operations should verify the identity of the cell batches prior to their use in experiments. A quality-control scheme similar to the one proposed by Festing for animal breeders might work for the larger facilities, but the multitude of small laboratories maintaining one or two lines would probably find it expensive. Their levy charges, however, would likely be low, since the risk of contamination is much lower in facilities where only a few culture lines have to be kept separate. Such a scheme should, perhaps, be regarded as an insurance policy for research quality, similar to the professional malpractice insurance routinely bought by lawyers and most doctors.

An additional approach could be modelled on the practice of chemical manufacturers. They routinely specify the percentage purity of each of their products; the purer the chemicals the more they cost. Perhaps a similar strategy would be suitable for animal suppliers. Researchers dependent on absolute genetic homogeneity could opt to purchase the purest strains (possibly those tested by outside laboratories), while those with less stringent requirements could buy less expensive stock.

Similarly, it should be the supplier's responsibility to warn users when quality-control measures have failed, just as auto manufacturers warn purchasers of potential faults. Since the supplier is the only body with a list of all the users of the dubious product, the simplest and most effective method of notification is for them to send letters to recent customers, and to append warnings to supplies or advertisements. (The notification procedure undertaken by Kahan in the scientific literature was slow, and not guaranteed to reach those Charles River customers who stood to be most seriously affected by Kahan's revelations.)

Individual investigators whose published work involved questionable material clearly bear the responsibility of warning their readers that the results may not be reproducible. Journals should encourage this practice, as they share with the authors and the suppliers an obligation to rectify any misleading information they helped to disseminate.[32]

The final issue of compensation for research time lost as a result of reliance on contaminated material will probably be decided by the courts. Legal precedents already exist for analogous cases alleging fraud, negligence, and breach of warranty. They indicate that suppliers who fail to guarantee that their product lives up to its advertised qualities will have to recognize the purchaser's losses.

The vulnerability of modern scientific investigations to the inadvertent contamination of research materials indicates that quality control must be an important priority for scientific researchers. Science education should stress the fragility of the data produced by modern science, the potential for error in research, and the many ways in which the meaning of experiments and simple observations can be modified by false assumptions regarding the materials or methods used.

Even more pressing, though, is the recognition that these same problems, endemic in scientific investigations, provide opportunities for deliberate manipulation of experimental work. Using carefully chosen assumptions or a method that will give rise to certain kinds of results, investigators can unwittingly or even consciously steer the direction of their work to favour the conclusions they prefer. The operation of such scientific choices to promote various interests is investigated in Chapter Three.

Chapter Three
SCIENTISTS FOOL THEMSELVES

[T]he unpalatable possibility must remain that even professional scientists may have more in common with other professional people – businessmen and politicians, for example – than they would like to think. Deliberate fraud in an academic setting is un-doubtedly rare ... but who could say the same of, say, unthinking or unconscious self-deception? What happens if a person finds some unexpected result in the laboratory, publishes it, wins all kinds of acclaim – and then finds that his result cannot be repeated? To say that he should instantly recant is to ask a lot of flesh and blood.

-- Editorial: "Is Science Really a Pack of Lies?"
Nature 303 (1983), p. 362

As this *Nature* editorial notes, we ask scientists to perform a virtually impossible feat – to remain resolutely sceptical about the theories they care most passionately about. We shouldn't be surprised if few scientists can live up to this ideal of behaviour. The examples in the last chapter show how even the most open-minded investigators, who set out with few preconceptions about what they will find, will inevitably encounter disappointments along the way. How much more troubling, then, are the scientific setbacks that face researchers who know exactly what they want or expect to find. The danger is that unwished-for evidence may be too disturbing to confront; consciously or unconsciously, scientists may interpret and even design their work to produce data that will confirm their cherished hypotheses.

Most scientists can't help but develop loyalties and become committed to their theories, their field of study, their colleagues, and their departments. Team spirit is encouraged, for effective collaboration is probably the most productive form of scientific enterprise. However, team spirit can become peer pressure to conform to the team's ideas, preconceptions, or values. A multitude of opportunities arise for directed research results, findings moulded to coincide with the researchers' hopes, values, and preconceptions. In a rare acknowledgement of such over-enthusiasm, Robert J. Gullis, a former biochemist at a Max Planck Institute, admitted inventing data to support favoured theories: "I wish to disclose the fact that papers published in several journals with myself as principal author are not reliable. The curves and values published are mere figments of my imagination, and during my short research career I published my hypotheses rather than my experimental results. The reason was that I was so convinced of my ideas that I simply put them down on paper; it was not because of the tremendous importance of published papers to the career of a scientist."[1] This is a very extreme example: the complete fabrication of imaginary data is, doubtless, a rare exception; but the pursuit of desired scientific results, through the choice of experimental methods, analytical techniques, interpretations, or reporting formats is probably common. To the extent that investigators' assumptions colour their interpretation of results, unwelcome conclusions, at least, can often be avoided.

Of course, it is almost impossible to tell whether self-deception is unconscious or if there has been a deliberate attempt to manipulate research results. At what point does speculation become conviction and supporting evidence become proof? Where is the line between scientists' eager acceptance of favourable results and designing research to produce desired data? When does naiveté stop and corruption begin? No one, apart from the individual scientists themselves, can know their hearts and minds – and even they may be blinded by their beliefs. For unconscious self-deception is probably the rule in science, not the exception, as most scientists would like to believe.

Two environmental controversies, over Love Canal and over the health effects of lead in gasoline, dramatically reveal what happens when scientists develop passionate allegiances to their data and their interpretations of them. Particularly in such public controversies, researchers have difficulty preventing their political and social sympathies from influencing their research. The process of job selection, and scientists' choice of financial supporters, colleagues, and research projects will both reflect and mould their academic and, perhaps, even political sympathies. It can be expected that scientists' findings in topical matters will usually support the policy positions advocated by their colleagues and benefactors.

Scientific investigators generally believe that their research results are solid, "objective" numbers, not subject to personal views. But figures alone can create an impression of solidity and "objectivity" that hides the uncertain or dubious assumptions on which they are based. Politicians, the media, and often members of the public want science to provide conclusive, precise answers to public-policy questions. Often such evidence doesn't exist, and attempts to create it merely cloak researchers' or statisticians' assumptions in otherwise meaningless numbers.[2] When data do exist, the scientific debate often shifts from the validity of the underlying assumptions to the appropriateness of the methodology and the analysis of the results. In some cases, like the controversy over environmental contamination at Love Canal, different decisions concerning the design of the investigation have resulted in inconsistent sets of results. In other situations, like the extended wrangling over the connection between lead in gasoline and lead in people's blood, the data were universally accepted; it was their analysis and interpretation that were hotly disputed.

The environmental controversy at Love Canal erupted in 1978, when the New York State Health Department identified more than eighty chemicals in the Love Canal chemical-waste site, including eight potentially carcinogenic compounds found in the air of Love Canal homes.[3] These houses were part of a suburban subdivision, which included a primary school. The development had been built directly over and around Hooker Chemical and Plastic Company's former waste-disposal lagoon, outside Buffalo, New York.

The events of the summer and autumn of 1978 followed what has now become a familiar sequence. The commissioner of health declared a health emergency on the grounds that "a review of all the available evidence ... has convinced me of the existence of a great and imminent peril to the health of the general public residing at or near the said site as a result of exposure to toxic substances emanating from such site."[4] He also made several recommendations that frightened the residents of Love Canal: they were advised to stay out of their basements, and to refrain from eating their own garden produce. Pregnant women and children under two were encouraged to move as soon as possible; and the local school, built on the original canal site, was temporarily closed. The State Health Department hurriedly embarked on a major health survey of the area, including blood testing, lengthy questionnaires, and a survey of residents' medical records.

The townspeople were understandably alarmed by these warnings from the State Health Department, an agency not known for overstating or exaggerating health risks. Eager to help find out how they were being affected by the chemicals, members of the Love Canal Homeowners' Association assisted in the health study. Association volunteers telephoned their neighbours to urge them to complete their questionnaires and medical-record release forms, and to note the medical problems each had experienced. In this way, the association amassed a sizeable body of anecdotal information on disease in the area.

Lois Gibbs, the leader of the Homeowners' Association, was certain that she and her family were suffering from the chemical emissions. She was keen to move out of the area, mainly to protect her son, who had developed several asthmatic symptoms and had had convulsions when he first went to the local school. After several months of community organizing and frustrating dealings with the State Health Department, she was losing her patience.

Late one night, she hit on the idea of marking the information reported to association volunteers on a street map, with pins to indicate homes with medical problems. As she worked on this project she thought she discerned a pattern of disease outbreaks concentrated in certain areas and along narrow paths. Might this physical arrangement coincide with the old drainage ditches that used to cut through the neighbourhood?

Excited by this possibility, Ms. Gibbs showed her map to the state official in charge of the Health Department study and to Beverley Paigen, a scientist studying susceptibility to environmental toxins at a nearby state health-research institute. Dr. Paigen eagerly seized on this research opportunity and devised a study to test Gibbs's hypothesis. First, she designed a simple telephone questionnaire, which was carried out by Homeowners' Association volunteers. In total, 1140 people in 850 families were surveyed; when the results were mapped, the clusters of illness seemed to follow the former drainage beds. These clay beds had been filled in with rubble when the subdivision was built, and Dr. Paigen speculated that depressions in the clay might still provide a conduit for toxic seepage from the canal. If the toxic chemicals emanating from Hooker's old waste site were concentrated in certain areas and paths through the neighbourhood, then illnesses might be as well. To check whether her first impressions, based on the mapped data, were justified, Dr. Paigen divided the suburb into "wet" homes, built on drainage beds, and the other, "dry" homes in the area. She then compared the incidence of disease in these two groups.

Her results were striking. Women who lived in wet homes had an exceptionally high rate of repeated miscarriages, and three times as many miscarriages as those in dry homes. Birth defects, asthma, urinary-system disease, and admissions to mental hospitals were also several times more frequent among families living in wet areas.

The results of this study clearly supported Lois Gibbs's hypothesis. They also had important implications for how this health emergency should be dealt with. Dr. Paigen's finding implied that any evacuation scheme should include residents of the "wet" houses, no matter how far they were from the canal, the original source of the toxic chemicals. The plan devised by the New York State Health Department was, however, rather different. The state was offering to buy up an "inner ring" of houses immediately adjacent to the canal site; families living beyond a specified distance from the site were, initially, expected to stay on.

Dr. Paigen realized that more research was needed to follow up her simple study carried out on a shoestring budget. Her investigation had not set out to provide academically watertight data, and she hoped that the New York State Health Department would at least respond with a more rigorous, comprehensive inquiry. But the state scientists were not keen to pursue Dr. Paigen's hypothesis. Her results were in striking contrast to their own, which had been distilled from much more meticulous examination of medical tests, records, and detailed questionnaires. When they announced their first findings the State Health Department conceded that the wet areas of the neighbourhood did seem to be chemically contaminated, but they had found no evidence that this contamination was associated with increased illness. Their initial analysis supported the conclusion that the miscarriages at Love Canal were not related to chemical exposure.

How could two studies on the same basic question, conducted by reputable scientists, produce such contrary conclusions? To find out, we have to compare the way the investigations were designed and analysed. On the surface, the Health Department investigation seems similar to, albeit much more thorough than, Dr. Paigen's effort; but its basic assumptions reflected a fundamentally different scientific approach and social outlook.

The Health Department study was planned to test the very reasonable hypothesis that chemicals were seeping away from Love Canal at a more or less even rate. Logically, they compared the miscarriage rate on different streets, at increasing distances from the canal. They found that this rate was not higher in women closer to the original source of the toxic chemicals, the canal under the school and adjacent playground. Moreover, when they compared the overall miscarriage frequency in the neighbourhood with one reported in the literature, they found a comparable incidence. (This comparison was criticized on the grounds that the reported finding presented a miscarriage rate for poor women, most of whom had already had children with birth defects; the rate for such women would have been higher than average.) Health Department scientists were convinced that the residents of the "outer ring" of Love Canal, houses that were not being bought up by the state, were not at a significantly greater health risk than the rest of the American population.

When Lois Gibbs then released Dr. Paigen's preliminary results to the press, Dr. Vianna, the scientist in charge of the New York State Health Department study, asserted that her findings were "totally, absolutely, and emphatically incorrect".[5] The study was dismissed as "information collected by housewives – that is, useless".[6]

Several months later, however, after the Health Department investigated the significance of the old drainage ditches, Dr. Axelrod, the state health commissioner, admitted that miscarriages and low birth weights were indeed more prevalent in historically wet homes.[7] In 1980, this conclusion was stated very tentatively in a State Health Department report on Love Canal: "With a great deal of caution we can speculate on the implications of our findings with regard to the possible modes of exposure to Canal chemicals.... Our observations appear to be most consistent with hypothetical exposure related to historic low lying [wet] areas."[8]

Why was this conclusion embraced only with reluctance by State Health Department officials? In this story, as in others, we appear to have, on the one hand, a cautious approach to scientific analysis and a reticence to leap to dramatic conclusions, and, on the other hand, a bold but possibly ill-considered scientific model. In fact, however, both camps have a common ground: an unacknowledged extension of political activity into the scientific realm, in supporting favoured public-health policies with relevant epidemiological evidence. Both Dr. Paigen and the Health Department researchers are trained public-health scientists. They are all competent scientists, of personal and professional integrity. However, an examination of the contexts for the two studies shows that their investigations were based on rather different perspectives on the issues raised at Love Canal.

The Health Department officials were concerned, but they were also committed to defending their actions with solid, indisputable scientific evidence. Their thorough, rigorous study, based on verifiable medical records and blood tests, was intended to provide exactly this. Their insistence on what they termed a "conservative" scientific approach was also geared to the inevitable political battle for funding any actions indicated by their results. And they were all too aware of how much money might be required: according to a State Health Department estimate, there were about a hundred chemical dump sites in Niagara County alone, and Love Canal was by no means the largest or most dangerous. The projected cost of remedial action at Love Canal was estimated in 1981 to total more than sixty million dollars.[9] In order for such action to be considered State Health Department officials would have to prove not only that there were very definite health problems, but also that they were indisputably connected to the presence of the Love Canal chemicals. The usual scientific technique for establishing such relationships is to prove that they are statistically significant. In other words, State Health Department scientists had to show that the chance of their conclusions being wrong was less than 5 per cent.

Dr. Paigen, however, was convinced from early on that chemical contamination in the area was responsible for health problems.[10] She suffered from asthma, nausea, upset stomach, and severe migraine headaches when she worked in the area. The effects were so debilitating that she eventually had to limit the time she spent there. Her belief that her own ill health resulted from exposure to Love Canal chemicals led her to reject the Health Department's original conclusions that the chemicals weren't causing undue problems. Her background as an environmental advocate who had worked with broadly based citizens' groups in the past allowed her to welcome the idea put forward by Lois Gibbs, even though it wasn't backed up by any conclusive scientific evidence. Moreover, her close involvement with local residents and their gratitude to her reinforced her concern for their health: she was eager to detect any unusual health problems, even if they could not be conclusively related to the chemical seepage using the rigorous statistical and methodological standards of traditional scientific publications. She felt that the public-health risk of overlooking or underestimating actual disease related to chemical exposure was much more dangerous than the community anxiety that might be raised by groundless fears.[11] And Paigen was prepared to lose peer acclaim by raising these issues. Nicholas Ashford, director of the Massachusetts Institute of Technology's Center for Public Policy Alternatives, illustrated this approach as asking not "can you publish it in the *New England Journal of Medicine*, but would you let your daughter work with that chemical?"[12]

Any immediate response to Dr. Paigen's conclusions by the New York State Health Department would have cost a great deal of money, and would have set a dangerous precedent for a state with more than nine hundred chemical dumps – many of them toxic – within its boundaries.[13] So it would have made tactical sense for Health Department scientists to err on the side of scientific fastidiousness and to avoid raising hopes or fears that they would be unable to act on or allay. Eventually, the New York State Health Department did test Lois Gibbs's hypothesis by checking the incidence of low birth weights. They reported: "A statistically significant excess was found in the historic swale (wet) area from 1940 through 1953, the period when various chemicals were dumped in this disposal site."[14] The State of

New York, assisted by federal funding, did take several steps to deal with the Love Canal problem. The government closed the local school permanently, and bought more than seven hundred houses, enabling the residents to move elsewhere. The deserted neighbourhood remains a testament to our short-sighted and cavalier treatment of dangerous chemicals. This dispute illustrates graphically how the standard of proof, the criterion for accepting evidence as true, seems to be adjusted to conform to the scientist's preferred conclusions.[15] Welcome results are accepted despite their deficiencies, while data that call into question a cherished research hypothesis are subjected to intense critical scrutiny.

Examples of environmental controversy like the one at Love Canal bitterly inflamed by contrary scientific findings, can be found around the world. In Vallecitos, California, engineers debated for years the risks of locating a nuclear-power plant near a geological formation that could be a fault or a relic of an old landslide before it was ultimately decided to start the plant up.[16] Scientists continue to wrangle in public over the amount of lead that can be safely tolerated in children's blood,[17] and the decision pending on permitting food irradiation in Canada is surrounded by heated scientific as well as public-policy debate.[18] In these and other similar controversies, scientists on opposite sides of a political fence collected different data to support their divergent preconceptions about what was acceptable for human health and the environment, and what they were willing to pay to achieve it.

By and large it is easier to support different points of view by using different data, even when, as in the Love Canal case, the scientists are supposedly trying to answer the same question. However, in the continuing controversy over the contribution of lead in gasoline to blood-lead levels and the attendant dangers, the antagonists used the same enormous data set to promote contrary conclusions. This story helps to illustrate why science will often fail to provide the precise, concrete information that politicians and regulatory officials need.[19] Even when the data are universally accepted, the methods used to analyse and interpret them are often so varied that widely divergent conclusions can emerge from a single set of scientific results.

Lead and its effect on human health arouses passionate debate.[20] Humans can tolerate remarkably high levels of lead: no obvious ill effects are manifest even when a person's blood contains up to one-quarter of the lethal lead dose. The average urban Ontario child's blood includes twelve micrograms of lead per decilitre (μg/dl).[21] Twenty μg/dl is the level at which Health and Welfare Canada says adverse health effects can occur in children – 80 to 100 μg/dl have been known to cause convulsions and even death in children.[22] When the concentration of 20 μg/dl blood lead is detected in children, the Toronto Health Department becomes concerned about damage to public health. Thus, many Canadian children are contaminated with lead at more than half the level that can cause health problems. In fact, the most recent data have led to a "very strong international consensus that blood lead levels between 10-15 μg/dl [are] of serious concern".[23] For many other damaging chemicals, an allowable level of one one-hundredth of the amount proven to cause harm is a generally recognized safety margin.[24]

Clearly, the measure of protection provided in the case of lead is very much weaker, even considering our unusual tolerance of relatively high concentrations of lead in our blood. As a result, lobby groups in many countries press for elimination of lead in automobile gasoline, the source of about 63 per cent of atmospheric lead pollution.[25] The companies producing the lead additives are reluctant to lose the massive profits from these sales and, understandably, oppose any reductions in gasoline lead.

In March 1982 the Centers for Disease Control (CDC) in Atlanta, Georgia, published a graph showing a simultaneous and similar decline in lead in gasoline (due to the introduction of unleaded gasoline) and in the lead levels in Americans' blood between 1976 and 1980.[26] The data seemed to imply that lead in gas was at least partly responsible for the high lead levels in human blood. Dr. James Pirkle, a physical chemist and medical doctor at the CDC, finished his paper with the words: "These studies strongly indicate that removing lead from gasoline would significantly reduce lead exposure for adults, children, and infants awaiting birth."[27]

The laboratory results (taken from the Second National Health and Nutrition Examination Survey, NHANES II) were passed on to the federal Environmental Protection Agency (EPA). Joel Schwartz, an EPA scientist who had been doing refinery cost analysis of lead in gasoline to estimate the economic benefits of lead retention (in preparation for the possible deregulation of lead in gasoline), drew even more forceful conclusions from his re-analysis of the same data: "Therefore, we conclude that there is a strong causal relationship between gasoline lead and blood lead, and that gasoline lead probably contributed more than half of the observed blood lead in the U.S. population in the late 1970s."[28]

The CDC data also became available to the public and, eventually, the Ethyl and E. I. Du Pont de Nemours & Co., Inc. (DuPont) corporations undertook their own re-analyses of the figures. These industrial giants, both with heavy financial stakes in leaded gas, came up with strikingly different conclusions that gas lead is *not* the primary contributor to blood lead. Ethyl Corporation scientists found that the "Statistical analyses of the NHANES II data indicate that the major part of the decrease in blood lead during the survey period *was not due* to the decline in gasoline lead usage" (emphasis added).[29] DuPont scientists, similarly, pointed out that changes in the population tested over the four years, and not the decline in gasoline lead, were the most important factors: "*Changing demographics of the subject groups* at the 64 sites sampled during NHANES II *accounts for over half* the apparent 5.61 µg/dl *blood lead decrease* from 1976 to 1980" (emphasis added).[30]

Once more, researchers and statisticians who are considered to be good scientists arrived at totally divergent conclusions. Certainly it wasn't a question of incompetence leading to false analysis; but the scientists clearly fell into two polarized camps, and each side had strong convictions about what the data meant. Whether the researchers had these beliefs at the outset or developed them along the way is impossible to know. What we can learn by charting the history of this controversy is how the biases of each protagonist became manifest in their apparently "technical" and "objective" scientific work.

NHANES II was designed by the National Center for Health Statistics for the USFDA (United States Food and Drug Administration) as a cross-sectional study of the U.S. population, to evaluate their general health and nutrition during the years between February 1976 and February 1980. It was a multimillion-dollar effort to compile one of the most comprehensive sets of epidemiological data ever assembled. Many different blood components were assessed, for nearly ten thousand blood samples. These samples were analysed for lead at the Centers for Disease Control in Atlanta, primarily so that the federal Department of Agriculture could evaluate the importance of dietary sources of lead.

A disturbing trend in the blood-lead results was detected: it appeared that some sort of contamination might have been present early on in the study. Researchers were understandably alarmed, and Dr. James Pirkle was called in to check on the quality-control procedures.

Dr. Pirkle expected his task to take a couple of weeks, for standardized samples containing a measured quantity of lead had been interspersed with the real samples taken by the scientists carrying out the study. If any of the real samples had been contaminated with extra lead from the laboratory where they were analysed, the same contamination would also show up as elevated lead levels in these standardized samples. But when Dr. Pirkle systematically re-examined them, the standardized samples turned out to be clean: they contained only the uniform amount of lead that had initially been apportioned to them. Nor could Dr. Pirkle fault the analytical procedures used. On the evidence that the standardized samples had been analysed correctly Dr. Pirkle concluded that all of the other lead samples must have been analysed correctly as well.

With the most obvious source of error – laboratory contamination – eliminated, what could have caused the change in blood-lead levels? Dr. Pirkle decided to examine all the characteristics of the people sampled that might have affected their blood-lead levels. With Dr. Lee Annest of the National Center for Health Statistics (NCHS), he embarked on a thorough statistical investigation of the data. They ran extensive computer programs to check on the association of the blood-lead values with race, sex, age, season, income, region of the country, and degree of urbanization of the individuals sampled. But none of these factors could explain the consistent decrease in blood-lead levels over time. At last, Dr. Pirkle and his colleagues focused their attention on human lead exposure, looking first at food and paint, and then at gasoline. Finally they turned to the increased use of unleaded gas, which had caused a decline in the total amount of lead added to gas by the refining industry.

Total gasoline lead levels, the very last factor to be suspected of contributing to the blood-lead decline, showed a striking association with blood-lead, even when each of the subpopulations, characterized by factors such as age, sex, and racial group,[31] was examined on its own. Apparent seasonal variations in blood-lead were also related to differences in the average amount of gasoline lead used. While these associations did not *prove* that the decline in gas lead caused the reduction in blood-lead levels, the two trends were remarkably similar.

The strength of a relationship like this can be measured using a statistical indicator which has a value between zero and one. When the two trends being examined exactly correspond, this indicator, called a correlation coefficient, registers "1"; when the two trends are entirely independent or unrelated it registers "0"; and when they are more or less strongly associated, this is reflected by a correlation coefficient with a value between zero and one, which is closer to one for a stronger association and closer to zero when the reverse is true. According to Dr. Pirkle, his results were unequivocal: "When the correlation of .95 came out, I was awestruck ... it made a gigantic impression."[32] Dr. Pirkle's report, including the now famous graph showing the correlation between the decline in blood-lead levels and total lead in gas, was published in the CDC's "Weekly Mortality and Morbidity Report" on March 19, 1982.

Joel Schwartz, the EPA scientist who eventually became deeply embroiled in this debate, was originally intrigued by the potential environmental costs involved in increasing the amount of lead in gas, which was being considered as part of a move to deregulate the industry. To ascertain whether the connection between lead in gas and lead in blood was as strong as the CDC scientists asserted, he decided to re-analyse the data for himself. Assisted by an independent consulting group, he carried out slightly different calculations based on the CDC data and reached similarly categorical conclusions regarding the link between gas lead and blood lead.

Ethyl and DuPont then re-examined the data, embarking on their own, rather different analyses, which came to the conclusions quoted above.[33] The DuPont analysis first described each sampled individual in terms of fifteen variables related to personal, residential, geographical, and temporal qualities, such as age, race, income, and location. The DuPont researchers then adjusted each sample lead value to cancel out the influence of each of these characteristics. On this basis, they concluded that the resulting "corrected" blood-lead values did not exhibit as pronounced a decline over the four years of the study as the "raw" or uncorrected data did.

Instead of directly comparing the lead in gas to that appearing in human blood at the time, they multiplied the amount of gasoline lead used in each state by the proportion of its population in each county, and then divided by the county area to get an average gasoline lead exposure for each person in each county. By combining the gas-lead units with this measure of county population density they effectively concealed any impact of the change in the amount of lead in the gas over the four years on people's blood-lead levels. Unlike other gas-lead measurements, DuPont's showed little change over time. DuPont scientists concluded that the date (or time) of blood analysis was the most important determinant of the amount of lead in a person's blood, and they made what seems to be a groundless assumption, that this time factor represented all the changes in the non-gasoline sources of blood lead.[34]

Ethyl scientists also emphasized the variation in the sampled population over time and claimed that personal characteristics were very important in explaining individual blood-lead levels. Like DuPont, Ethyl examined local use of lead in gas; the measure they chose reflected population density while disguising the important national trends in lead added to gasoline. [35] Ethyl scientists actually *assumed* their conclusion: "that gasoline lead has the same per cent contribution to blood lead relative to other sources in 1980 as in 1976", and so it was inevitable that their lead-use measure, like DuPont's, masked the relation of the blood-lead data to gasoline lead. [36]

In the face of these contrary analyses, the EPA convened a panel of reputable statisticians to review the conflicting interpretations of the blood-lead data. The report of this panel, the NHANES II Time Trend Analysis Review Group, strongly supported the EPA and CDC analyses, while damning the efforts of Ethyl and DuPont scientists with very faint praise: "The CDC and ICF analyses provide strong evidence that gasoline lead is a major contributor to the decline in blood lead over the period of the NHANES study. DuPont stressed the limitations of statistical theory and methods as tools for assessing causal relationships.... The Review Group finds that the Ethyl analyses contribute little to understanding the association between blood lead and gasoline lead because the variables adopted to represent lead exposure are deemed inappropriate." [37] A year later, in a pre-election move, the EPA proposed a new regulation that reduced lead in gas by 91 per cent as of January 1986. Supporting documents refer repeatedly to the CDC and EPA analyses of the NHANES data, [38] and in a press conference accompanying this announcement William D. Ruckelshaus, the EPA administrator, reflected the current consensus of expert opinion when he asserted: "We know that there is a direct relationship between lead in gasoline and the amount of lead in human blood." [39]

In this example it is easy to see the vested interests of all the institutions involved. Among their other corporate activities, Ethyl and DuPont process and market lead. Ethyl's 1983 *Annual Report* states that the company "set new records in 1983, as net income surpassed the $100-million mark for the first time....

Lead antiknock compounds [which are added to gasoline to make it burn more smoothly at a higher compression] represented about 26 percent of Ethyl's operating profit in 1983."[40]

Dr. Donald R. Lynam, director of air conservation for Ethyl and first author of their paper on lead, is listed as a senior officer of the company. He's been following the lead literature for more than sixteen years, and has previously published on the subject of lead. He feels that the review committee was hand-picked by the EPA, presumably to support their regulatory bias.[41]

Similarly, Dr. Pfeifer, who works in the Engineering Department at DuPont, started out with a perspective critical of the CDC and EPA analyses. He admits that his concern with this issue springs from its relevance to the business he's in and that his findings are used to justify DuPont's policy towards the government stance on lead. He is passionately convinced that gasoline lead is only a small contributor to blood lead, making the astonishing testimonial that, "If you were to ask God what is the true decline in blood lead attributable to gas lead, He would say much less than the five micrograms found by the CDC and EPA."[42]

DuPont and Ethyl have an obvious financial interest in a weak relationship between gas lead and blood lead. Ethyl documents the declining contribution of lead antiknocks to its profits as regulation of lead has become more stringent.[43] Millions of dollars of profit hinge on whether the regulators are convinced that the gas-lead/blood-lead association is strong enough to call for more stringent control. It would be difficult for company scientists to interpret the data in ways that would conflict with the interests of the company that employs them. Such interpretations might also lead to the painful admission that the entire premise of their research work was misconceived; maintaining such an attitude in the face of corporate hostility would be traumatic. On the other hand, company scientists have accused the scientists who advocate stricter lead regulation of systematic bias in favour of a relationship between blood lead and gas lead.[44]

The case for systematic and preconceived bias is more difficult to establish for government and other scientists whose work links gas lead and blood lead, despite efforts to generally discredit scientific work supporting more stringent environmental controls.[45] While there is no question that the EPA has environmental preservation as its mandate, it appears that neither EPA nor CDC investigators had a direct interest in proving a strong relationship between gas- and blood-lead levels. James Pirkle's initial investigation was based on the premise that the apparent decline in blood-lead levels during the period of the NHANES study was a misleading result of poor laboratory procedures. It was only after his thorough research on quality control proved otherwise that the alternative – namely, that the decline in blood lead was a real phenomenon – was considered. Joel Schwartz, at the EPA, had been working on economic studies supporting the deregulation of lead in gasoline; he switched to work on blood lead only after the NHANES II data had been analysed by the CDC. Then, not content to rely on their evaluation, he reassessed the data, using slightly different methods to ascertain their import.

None the less, regardless of whether they started out with definite opinions on the subject, it is clear that scientists working on the blood-lead data developed strong allegiances to their stated viewpoints. As Dr. Richard Royall, a member of the government's NHANES II Time Trend Analysis Review Group asserted, "The scientists [involved] believe what they want to believe.... At the moment [they] make some public statement and get identified with a point of view, it's hard to admit they're wrong."[46]

In fact, as soon as scientists have a personal identification with a point of view, even if it isn't publicly declared, their scientific conduct may be swayed by it. The social and career pressures to conform in action – and even in thought – to the views and attitudes of colleagues are enormous, especially when the options for dissenters are extremely unattractive. Whistle-blowers are not rewarded in corporate, political, or academic contexts; without vigorous regulatory protection those raising a public objection to a prevalent scientific practice are likely to jeopardize, if not destroy, their future work prospects.[47]

Scientists' personal experiences also contribute to their view of science and their use of it. June Goodfield, in *An Imagined World*,[48] and Evelyn Fox Keller, in her biography of the pioneering geneticist Barbara McClintock, *A Feeling for the Organism*,[49] try to relate the personal and professional experiences of one scientist to her discoveries.

In a review of 345 studies on the effect of experimenters' expectations on their research, David Rosenthal and Donald B. Rubin concluded that there is a definite tendency for investigators to shape the experimental results they obtain.[50] Clinical psychologists have also demonstrated that people generally find existing evidence in support of previously held views more convincing than contrary data. (For example, a study on Stanford University students published in 1979 concluded that when individuals with strong opinions on capital punishment were given contradictory evidence regarding its effectiveness as a deterrent, the evidence strengthened the beliefs already held.)[51] That is, people tend not to be swayed by evidence that does not support their own convictions; exposure to the *same* information merely polarizes the audience!

The implication of such studies is that scientists' research evidence is unlikely to be unequivocal; even precise, numerical measurements will probably be interpreted, as far as possible, to support the perspective of the scientist examining it. Literature in the history and sociology of science confirms this view; in a recent article, the historian of science Thomas Kuhn asserts that "nature undoubtedly responds to the theoretical predispositions with which she is approached by the measuring scientist."[52] Brian Wynne, another historian and sociologist of science, also finds that, in the case of late Victorian physics, "the concepts and principles of a science were developed and sustained not only (or perhaps not even) for their technical value, but very much also for their social value."[53]

These biographical, historical, and psychological investigations, like those associated with Love Canal and lead in gas, make it abundantly clear that the way the scientific problem is posed, the way the investigation is designed, the way the data are collected, analysed, interpreted, and presented are all a function of what scientists want to prove. Why are they doing this research, after all? If future scientists are to become aware of this fact, science educators should no longer pretend that scientific data can easily be separated from the scientist who assembled them. Instead, recognition and anticipation of this sort of bias should become part of the operating ethic for all scientific researchers. Science courses should stress the inevitable intrusion of the investigator's expectations into research design, and practical techniques should be developed to make these influences explicit and to encourage scientists to act as their own devil's advocates. As the late Sir Peter Medawar, Nobel laureate and experimental pathologist, cautioned in his book *Advice to a Young Scientist*, researchers should be warned not to cling too doggedly to cherished theories:

> There is no certain way of telling in advance if the day-dreams of a lifetime dedicated to the pursuit of truth will carry a novice through the frustration of seeing experiments fail and of making the dismaying discovery that some of one's favorite ideas are groundless.
>
> Twice in my life I have spent two weary and scientifically profitless years seeking evidence to corroborate dearly loved hypotheses that later proved to be groundless; times such as these are hard for scientists – days of leaden gray skies bringing with them a miserable sense of oppression and inadequacy. It is my recollection of these bad times that accounts for the earnestness of my advice to young scientists that they should have more than one string to their bow and should be willing to take no for an answer if the evidence points that way.[54]

Politicians and regulators must also learn to expect of science only what it can realistically provide. It has become commonplace for judges, juries, and administrative tribunals to be faced with conflicting scientific evidence from credible scientists. When public, corporate, or government interests are at stake, as soon as one scientist makes a public statement asserting, for example, that a certain product is unsafe, another will immediately rise to deny the argument. Undisputed scientific evidence that categorically informs policy debates and is universally seen to give support to a particular position is bound to be the rare exception.

Science administrators, publishers, and funding agencies should also accommodate and even welcome advocacy research. Rather than vainly insisting on "objective" investigations, they should concern themselves with the expectations of those doing research work. They should insist that scientists declare their history of involvement with issues when making public or academic pronouncements on them. Most important, government agencies and tribunals should not fund only one study on any important problem. Recognizing the disparities in approach likely to be taken by different groups, funding bodies should actively support opposing groups of scientists in order to expose the weaknesses in the work of each.

Similarly, individual members of research teams should be encouraged by their sponsors to submit minority reports or alternate analyses of results.[55] This approach would give the politicians, regulators, media, and the interested public a broader and more useful basis for drawing their own conclusions regarding the social and political implications of scientific research. Particularly on matters of intense and urgent public interest, such as public-health and environmental issues, funding one umbrella group or a single government agency to conduct only one study in the interest of cost-saving must be seen as basically unsound.

None of this should be seen as aspersions cast on individual researchers or on the collective community of scientific investigators. Indeed, scientists probably try harder than almost any other group consciously to divorce their conclusions from their own personal role in their work: their proclaimed goal is to produce evidence that can be reproduced exactly by any other investigator. The field of statistics has been developed in order to make the evaluation of research results less capricious and more "objective". Yet it is inevitable that scientists, like everyone else, harbour personal aspirations which must affect their work. This interaction between science and the world in which it is conducted is a double-edged sword. While vested interests can clearly distort scientific work until it becomes useless, an awareness of the ultimate impact of scientific work on society and nature can make it more relevant, humane, and productive. In an endeavour in which conflicts and controversy often pave the way for new understanding,[56] criticism from passionate opponents clarifies and ultimately strengthens scientific work.

Chapter Four
PUBLIC KNOWLEDGE AND PRIVATE INTERESTS

The Canadian government is urging universities to transfer technology [to the private sector for commercial development] when possible. As a result, the government expects researchers and universities to move in a direction that will help implement this policy by adopting flexible rules for the use of government funding. McGill in this case appears to have embraced the government policy.... In the enthusiasm of attempting to effect the transfer of technology for the benefit of society, the Professors and McGill, the University failed to prevent a major conflict of interest situation.... [I]t was the fundamental conflict of interest between the Professors, their companies and the University that was at the root of all the problems.

-- A.K. Paterson, "A Report to the Principal of McGill University Concerning Events in the Department of Microbiology and Immunology from 1981 to 1983, Including Recommendations with respect to the Involvement of Academic Personnel in Commercial Activities", March 1984, pp. 82-83

The above is an excerpt from the 1984 report of Alex Paterson, a prominent lawyer, commissioned by McGill University in Montreal to investigate scientific conduct on campus. In an unprecedented move, McGill asked an individual from outside the academic community to look into the activities of faculty members, in this case Irving DeVoe, the former head of the Department of Microbiology, and Bruce Holbein, an associate professor in the same department.[1] The following account is based primarily on Paterson's findings.

While teaching and conducting research at the university, DeVoe and Holbein developed a new technique for removing metals from liquids. This invention promised great financial rewards; it had potential applications to the recovery of precious metals from mine tailings, the purification of toxic wastes, and the prevention or reduction of rust in nuclear reactors.

Eager to capitalize on their invention, DeVoe and Holbein applied for international patents on the process, and set up companies to pursue the commercial development of their new product. Rather than establishing and funding private research facilities, DeVoe-Holbein Canada Ltd. operated out of rented departmental research laboratories at McGill. Funding for company ventures was diverted from Dr. DeVoe's academic research grant from Canada's Natural Sciences and Engineering Research Council (NSERC), money earmarked for academic work that would profit the public at large. He borrowed $56,000 of departmental and grant funds to equip the company, and his wife, an experienced research assistant, was paid just over $1000 out of Holbein's NSERC research grant to do work which related, in part, to company research. Although the borrowed money was eventually repaid, had DeVoe asked NSERC's permission for the loan, it would have been denied.[2]

McGill researchers charged that this pursuit of corporate goals on university property was highly inappropriate, and the controversy created an unpleasant atmosphere in the microbiology department. New locks were fitted on the doors to the company laboratory, and university researchers were barred from it; company personnel had to keep company business secret. Access to company labs for use of equipment and supplies housed there was restricted to the company's office hours. Not surprisingly, atmosphere in the research area was tense and resentful.[3]

In the face of these allegations, Holbein and DeVoe were persuaded to take unpaid leaves of absence from McGill University, and by early 1985 both had resigned from their academic positions. At their plant in the Montreal suburb of Dorval they now work full time to develop, produce, and promote their new invention.

The DeVoe-Holbein affair shocked many people in the Canadian academic establishment. It raised issues of academic ethics and corporate responsibility that had, rather complacently, been avoidable in the past. It also forced the scientific community to confront the problems involved in the development of marketable technology, with all the temptations and frustrations that accompany corporate/academic collaboration. Finally, the perceived need for an outside investigator implied that the normal, largely unwritten code of upright academic behaviour was inadequate to deal with scientific business ventures on campus.

The failure of the traditional academic code of behaviour to address corporate science has been recognized for some time in the United States, where reduced government science funding and potentially lucrative advances in biotechnology have hastened the marriage of academic and corporate interests. In 1984 there were nearly three hundred small U.S. biotechnology companies like the one started by DeVoe and Holbein.[4] Many established firms have injected massive funding into university departments, under a variety of novel arrangements. Industrialist Edwin Whitehead, for example, invested $127 million in the Whitehead Institute at MIT to be managed in collaboration with the university's biology department. Massachusetts General Hospital has a department of molecular biology dubbed the "Hoechst department" after a $70 million deal to fund medical research was signed with that pharmaceutical/chemical company in 1980.[5] In all, industrial funding of academic research and development in the United States quadrupled between 1973 and 1983, increasing from $84 million to $370 million.[6] Such corporate/university contracts take various forms: Stanford Medical School and Syntex Pharmaceuticals set up a collective consulting arrangement whereby all faculty work eight days per year for Syntex in exchange for a lump sum donated to the department budget;[7] the Semiconductor Research Cooperative, supported by the major U.S. electronics corporations, dispenses funds to scientists to work on basic research in this area;[8] and a $23 million contract between Monsanto Corporation and Harvard University provides for twelve years of research support in exchange for exclusive patent rights to any discoveries concerning TAF (tumour angiogenesis factor), a controversial biological substance believed to regulate the growth of tumours.[9]

Canada has fewer such arrangements, but the financial pressures on universities and scientific researchers, government cutbacks in support of science, and the urge to create wealth and employment through industrial development in new fields such as biotechnology and microtechnology conspire to push industry and universities together. Private investment in academic research has increased; simultaneously, government-funded research in universities is being directed increasingly towards fields and problems likely to yield industrial rewards. A recent report on corporate sponsorship of academic research notes that at the University of Toronto, according to President George Connell, the current 3 or 4 per cent corporate contribution is increasing rapidly.[10]

The following Canadian examples of this *rapprochement* between academic and corporate interests illustrate a more general international trend. The private funding of research in Canada has grown from 31 per cent in 1973 to 46 per cent in 1984, and corporate officers and university presidents meet regularly to promote further collaboration.[11] University of Toronto vice-president David Nowlan recently urged businessmen to collaborate more with academics in research initiatives, and denied that the university culture interferes in any way with private-sector involvement in research.[12] The federal National Research Council of Canada funds the Connaught Medical Research Laboratories, to the tune of $8 million, to enable them to pursue biotechnology[13] and the Natural Sciences and Engineering Research Council (NSERC) has similarly identified targeted research on technology transfer from universities to industry as an area warranting more investment. "Research ancillaries", applied-research centres affiliated with universities but attracting support in the form of contracts for practical projects, have also become more attractive to university administrators.[14] Research on how to encourage the transfer of technology from universities to industry is specifically excluded from severe budget cuts to the Science Council of Canada, a federal science-policy development agency.[15] A further sign of the growing political interest in university/industry collaboration is the appointment on September 1, 1987, of Dr. G.A. Kenney-Wallace as chairman of the Science Council of Canada. Dr. Kenney-Wallace was formerly chairperson of the University of Toronto's Research Ancillaries Advisory Group and is an advocate of increased technology transfer. Recent commentaries by individual investigators and faculty associates show that Canadian scientists are increasingly apprehensive about the ultimate impact of corporate funding on their work and on the public benefits of university research.[16]

In Britain, the situation is rather different. While industrial funding of university research has been encouraged by the Thatcher administration, with an eye to corresponding cuts in government expenditures, industry is reluctant. In a report published by the Advisory Board for the Research Councils in May 1986, industrialists are described as seeing "no justification for an attempt by government to lay an additional burden upon the world of business in order to reduce public expenditures".[17]

In the United States, where academic/industrial collaboration is better established, the private sector already pays for more than half the total bill for research and development.[18] Knowledgeable observers speculate that continued increases in industrial funding will further depress federal science grants to those departments with industrial ties.[19]

While the exchange of desperately needed financial support for access to potentially profitable ideas and inventions clearly serves mutual needs, the convergence of academic and industrial activities in scientific-research laboratories is accompanied by a host of ethical, practical, legal, and political problems.[20] Many elements crucial to the general advancement of scientific endeavours may be lost along the way: the independent choice of research fields and problems, the openness and vitality of the laboratory atmosphere, the free exchange of views ideal for research advancement, and even the quality and integrity of the science itself, can be compromised by the intrusion of corporate motives on scientific conduct. Moreover, the public benefits to be earned from decades of government investment in university faculties will be seriously jeopardized if we fail to develop a clear, sound, and ethical relationship for universities and industry in the near future.

The most obvious pitfall of academic/industrial collaboration is direct conflict of interest. As the events in the McGill microbiology laboratories demonstrate, scientists who own or hold stock in companies that stand to profit from their research may be tempted to channel public grant money into private pursuits. It is also clear that there is often little to prevent a private concern from siphoning off interesting academic discoveries for commercial development before they are publicly discussed.[21] Indeed, such ventures are often actively encouraged by the university involved. Recent studies indicate that there may even be strong incentives for academics to have their work patented, for private profit.[22] It must be difficult, at times, to distinguish between applied research legitimately supported by government grants and commercial development work, which should be funded by the firm likely to profit from it. Scientists' obligations to the institutions and the public that funds their research grants can run counter to their private interests and commercial enterprises, especially in fields ripe for industrial exploitation.

Dr. David Baltimore, professor of biology at MIT and director of MIT's industry-funded Whitehead Institute, estimates the percentage of senior faculty in the major research universities with industrial connections at over 50 per cent.[23] Albert Meyerhoff, senior attorney for the U.S. Natural Resources Defense Council, reports that in the field of biotechnology virtually all the top scientists are connected with companies interested in this field.[24] Just how important commercial links can be to academic decisions is illustrated by the following anecdote, concerning Stanford University's hiring practices, related by Leon Wofsy, a prominent immunologist at the University of California in Berkeley:

> I once had, not long ago, a detailed description of how ... recruiting takes place from one of the recruiters. And this young faculty member whom he was talking about had to make a choice in taking his position on a faculty. And that choice was between two companies, which to affiliate with, each of which was headed by a different senior member of the faculty that he was associated with. So he finally made his choice, whereupon I was told by the other one, "Well, you know what that means. He won't be able to use my" – I won't mention which piece of equipment, because that would identify the person – "because, you know, we're business competitors now." And I said, "For crying out loud" – 'cause this is a friend of mine – "why in the world does a faculty member considering an appointment have to choose a company affiliation?" And the answer I got was, "Do you know how much my house cost?" It's a modest house, but it happens to be in the Stanford neighbourhood.[25]

This corrosion of the collegial relationships by business competition has other consequences, too. For example, openness can be stifled by concerns for corporate confidentiality. Scientists are unlikely to ask for help in interpreting new results if they fear that their colleagues might steal the march on any profitable developments arising from the research findings.[26]

Free discussion of ideas and active collaboration on research projects have been identified as key ingredients in the success of the British Medical Research Council's Laboratory of Molecular Biology, which has produced six Nobel prize winners. The lab canteen, where scientists take extended tea, coffee, and lunch breaks, is often the site of the most important exchanges of research ideas and theories. The laboratory itself is designed with "the minimum number of offices, and big labs, and as many shared facilities as possible to throw people together,"[27] according to Max Perutz, one of the founders of the unit and its first director. How unlike the locked doors and filing cabinets, and the hidden formulae and chemical recipes that characterize corporate facilities![28]

Taken to its logical conclusion, a proprietary interest in research results can cause important findings to become corporate secrets, hidden from the rest of the scientific community. Most industrial agreements to fund academic research include delays on disclosure, varying from several months to more than a year, so that company officials have time to establish the financial potential of the findings and get a head start on commercial development.[29] In the past, this pressure has even precluded publication,[30] and knowledge has effectively become private property.

In the case of industrial development of applied research, this may not be serious. But when basic research into the structure and operation of the natural world – such as work on recombinant DNA – is done for private gain, fundamental information remains secret, and future advances that depend on this knowledge will be delayed until the initial discoveries have been repeated in the public domain. This could set back important new discoveries, as underfunding of public science research slows the replication of the expensive work done in well-equipped private labs. In addition, application of this new secret knowledge to publicly worthwhile but non-commercial new developments could be severely delayed.

In his testimony to the U.S. House of Representatives Committee on Science and Technology in 1982, Albert Meyerhoff, speaking for the Natural Resources Defense Council, advocated mandatory public disclosure of scientists' financial interests in any businesses in a position to profit from government-funded research. He argued that this would restrain a scientist from depriving the public of benefits accruing from publicly funded research. The State of California has acted on his recommendations;[31] however, the California disclosure law applies only to state universities and to companies actually sponsoring research, but not to other business interests that might also benefit.[32] At McGill University, the Senate Committee, responding to the DeVoe-Holbein affair, developed weaker guidelines on this matter, which merely advise research staff to inform their department chairmen of all their consulting activities and any proprietary enterprises they may have.[33]

Ostensibly, academic research is funded by the public purse and non-profit agencies for the public's benefit. But the reality may be rather different. Even if the investigators involved hold no equity in private companies, they may be attracted to projects of benefit to corporations by the prospect of future industrial-research contracts. The universities, which take a percentage of all contract funds to maintain departmental facilities, are subject to the same pressures.

A real-life story illustrates the far-reaching effects of these pressures, which led to a judgement being entered in November 1987 against the University of California for failing to represent the public interest (as defined by the U.S. Congress in statutes governing agricultural research). The University of California at Davis received a relatively small grant from farm-equipment manufacturers to develop a mechanical tomato harvester. Following its production, academic departments bred new varieties of tomato – hard, square, and simultaneously ripening – that could be reaped efficiently by the new machine. These new developments drastically altered the nature of farming in California. They ushered in the wholesale mechanization of tomato culture, from the cultivation of tough tomatoes able to withstand a speed of thirteen miles/hour to the collection of twelve tons of harvested vegetables in an enormous plastic gondola that could be transferred to a truck for transport to the cannery.[34] The machine costs a small fortune (about US$200,000 in 1987) and most of the four thousand California tomato farmers, who had small holdings, couldn't afford to buy it. Nor could they sell their tomatoes, shipped in now-obsolete fifty-pound vegetable boxes.[35]

During the next decade the original four thousand tomato farmers were reduced to six hundred, and the average farm size increased tenfold as the smaller operators were bought out. Eighteen thousand farm labourers' jobs were lost to the machine, and tomato prices rose twice as fast as prices of other fruit and vegetables![36] Thus a small initial investment of private funds directed research towards the narrow interests of the large farms and equipment manufacturers, while directly harming the vast majority of farmers and the tomato-eating public.

The consequences of privately *owned* research can be even more drastic. For example, in the 1950s, Shell funded a researcher at the University of California at San Francisco to investigate the health effects of a pesticide they produced, called DBCP. That work revealed that the chemical caused rat testicles to atrophy, but this study was not made public for some time. Many years of DBCP use later, in 1977, the pesticide was banned in the United States because it caused infertility in the workers making it. The California State Health Department has noted that people using wells contaminated with DBCP have experienced higher than average rates of stomach cancer.[37] How many lives would have been saved if the original study had become public knowledge right away?

Perhaps more disturbing than the delay in publication of privately funded discoveries is the potential for corporate influence to bias the research itself. This was demonstrated in an alarming series of frauds and cover-ups permitting the licensing and use of over a hundred potentially lethal chemicals. A private U.S. laboratory called Industrial Biotest Laboratories (IBT) engaged in pre-licensing tests on new pesticides on behalf of the manufacturers. These tests were then submitted to the U.S. authorities to support the licensing applications. After the pesticides were licensed, distributed, and used, it was discovered that IBT and many similar facilities had manipulated and even invented data to "prove" that the pesticides were safe. These private testing laboratories, to encourage future contracts from the pesticide manufacturers, compromised not only the research, but the honesty and integrity of those reporting it. The test results were deliberately manipulated to promote the fortunes of the companies producing the chemicals and, by extension, those testing the chemicals. There are numerous examples of sloppiness and fabrication engaged in by these firms: for example, in one experiment the number of live animals increased after the chemicals were tested (presumably through the addition of animals in the middle of the experiment!); in another, mice born without eyes following testing were reported as "having difficulty in developing eye pigment".[38] Of forty-three IBT-tested chemicals, all or most tests on thirty-six have been judged invalid.[39]

Both the Canadian and U.S. governments investigated the IBT affair in 1977, when doubts were first cast on their data. Yet many of the pesticides approved on the basis of IBT's dubious data are still in use, and the public may be suffering long-term exposure to hazardous chemicals as a result.

The blatant corruption of those involved in the IBT scandal is an example of the most extreme consequence of corporate vested interests in scientific research. Less dramatic but probably more pervasive are the subtle changes in research direction that help industry more than the public, the unspoken pressure on graduate students to become involved in research that could benefit the private sector, and the gradual erosion of free and open lab practices through competition for profitable research results. Certainly, some privately funded research efforts – for example, the work in solid-state physics carried out at Bell Laboratories – have led to discoveries of fundamental scientific importance. Nevertheless, if control of scientific research becomes located in private hands, it will be shareholder profits, not public value or scientific advancement that will direct the bulk of research efforts.

Nicholas Ashford, director of the Center for Policy Alternatives at MIT, developed the following scenario:

> Imagine a bright young university biologist, interested in developing new *in vitro* tests for mutagenicity because of the possible social

benefits. He finds, however, that adequate funding for this project is not easily obtained. When his department chairman advises him that the university has entered into a multimillion-dollar contract with a corporate funder to produce patentable work in genetic engineering and he is invited to join the research team, he finds it difficult to decline.

Two important factors are at work here. On the one hand, of course, the academician has gone to where the money is. As a result, the direction of his research will be that desired by the commercial funder. He has also done that which pleases his department chairman. Spurred on by new-found academic activity, he may become less and less interested in returning to his original *in vitro* mutagenesis project. At some point, he may pursue bioengineering research, not out of obligation, but out of genuine interest. And then, even though this researcher is pursuing the research he wants to undertake, there may be a loss in academic freedom of the university, because his original research ideas are no longer being pursued there. Diversity is lessened.[40]

A similar concern is expressed by Auriol Stevens, Public Affairs Director of the British Committee of Vice-Chancellors, who fears that the erosion of the academic-tenure system in England, which guarantees the job security of university professors, will open the way to "bullying of an individual whose research may be thought to threaten important public or commercial interests".[41]

The price that we pay for moving research investment from pure to economically motivated research projects won't be known for decades. Eventually, the loss of breadth of new knowledge and the slowed pace of fundamental research findings may constrain fresh advances, and the emphasis on the commercial applications of results may discourage new theoretical syntheses of existing knowledge. For most major leaps in understanding come from new ways of describing and explaining existing knowledge, and the potential loss or delay of such important theoretical work because scientists are tempted by more financially rewarding investigations may be the most serious consequence of industrial/academic collaboration.

But there are other, more prosaic problems, too. Although industrial funding may be attractive when other sources of research funds are scarce, there is a distinct danger that when research does not quickly provide financial returns and patentable results, or when the industry feels an economic squeeze, academic-research contracts will not be renewed and faculty, students, and technicians will be left in the lurch. As John Servos at the Program in History and Philosophy of Science at Princeton University noted in his work on the history of MIT, industrial support makes a university department highly vulnerable to the economic climate:

> What had appeared initially as a natural and mutually beneficial alliance of businessmen and applied scientists revealed itself in the 1930s to be a temporary and unstable partnership. When confronted with a choice between retaining their own employees or subsidizing investigators at educational institutions, businessmen with near unanimity chose the former. Contributions to the support of applied research at MIT and other American educational institutions were marginal expenses to most businesses; when the need to economize became urgent, they were among the first costs to be cut. Those who had depended upon industrial support, at MIT as at other schools, were quick to suffer the consequences.[42]

If a scientist manages to secure a long-term source of industrial funding, then the question of intellectual independence must be raised. As reported above, in the case of the investigator working on the pesticide DBCP, or the IBT researchers doing pesticide-licensing studies, scientists' integrity and their obligations to the public interest can be compromised all too easily by their associations with industry. Corporate executives themselves recognize this, as revealed by this advice on how to "co-opt the experts" given by two university professors to business leaders:

> Regulatory policy is increasingly made with the participation of experts, especially academics. A regulated firm or industry should be prepared whenever possible to co-opt these experts. This is most effectively done by identifying the leading experts in each relevant field and hiring them as consultants or advisors, or giving them research grants and the like. This activity requires a modicum of finesse; it must not be too blatant, for the experts themselves must not recognize that they have lost their objectivity and freedom of action.[43]

To safeguard the public trust in academic researchers, several universities have now adopted policies on conflict of interest and publication delays.[44] One model set of guidelines was developed at the Mid Winter Immunology Conference at Asilomar, California, in January 1982:

> Faculty members who receive research support from any agency, public or private, should be willing to disclose the source and amount of personal income derived from private enterprises in areas related to their research.
> Peer review processes of granting agencies and journals which require access to confidential grant applications or the refereeing of manuscripts should not involve the participation of individuals who have a financial or commercial interest in the research areas under review.
> Acceptance by the university of support from any source for a faculty member's research should always be contingent on assurance of adequate provisions for peer review and the absence of conflicts of interest that compromise educational standards and commitment. A standing faculty committee (and I say faculty, not administration) should verify that acceptable standards of review

have been met and should, where there is doubt, initiate an appropriate *ad hoc* review procedure.[45]

Together, these requirements would no doubt improve the situation in which individual scientists have industrial ties: they would provide some assurance that findings made with public funds would not be siphoned off for private profit, and that graduate students would be somewhat insulated from unfair industrial pressures. Yet these measures would not affect corporate influence at the level of departmental or institutional policy. As David Noble, a researcher at the Smithsonian Institution in Washington, D.C. and a critic of recent academic/industrial links, points out, universities themselves have corporate alliances, and their record of self-policing is rather poor.[46]

More drastic measures would dramatically reduce the amount of industrial research performed in academic institutions: for example, academics might be required to choose between staying in the university using public funds and leaving it entirely if they work with industry;[47] they might be required to seek early publication of all research results; and universities might conduct impact studies on industrially funded academic projects. While there would be public benefits in securing the independence of academic scientists, the development and production of new technology originating in the university research laboratory would doubtless be delayed, and the public might be denied useful new products. Perhaps corporations that will ultimately benefit from basic scientific investigations could be required to pay a special research tax to support such work. A similar fund might usefully subsidize research of interest to public lobby groups that couldn't afford independent studies.[48] In any case, some method of ensuring an arm's-length relationship between those paying for scientific research and those carrying it out is essential. Guidelines on professional conflict of interest must be widely developed and applied, and corporate influence over a university department's research priorities should unquestionably be restricted.[49] Increasing government funding of basic research and incentives or tax schemes encouraging corporate research funding to be distributed by some independent agency would also help to secure adequate research support for publicly worthwhile projects. (This topic is discussed in detail in Chapter Seven.)

While the motive and means to use academic positions to further corporate goals are obvious, it is much more difficult to discern the prevalent if subtle ways that university scientists promote their careers at the expense of other scientists, or even by sacrificing the quality of their own research. By and large, this self-serving behaviour involves the kind of unconscious self-deception described in Chapter Three. At its most extreme, however, academic ambitions can lead to outright deception, the subject of the next chapter.

Chapter 5
THE ACADEMIC HUSTLE

If you are a successful scientist, you start off and do experiments with your own hands. You then progress to a point where you have a small group, three or four people working with you, all on related projects. You then probably progress up the ladder to a point where you have a large group, say 15 to 20 people even, doing experiments, and you have varying degrees of involvements in the individual experiments.... People will bring to you their data books and they will say: We got this result, we got that result – and you will discuss with them the interpretation, what experiment they should do next, and so on.... If someone in your laboratory says to you this is the experiment I did, these are the reagents I used, and this is the result I obtained, you are not likely to doubt it. Not unless there is something that looks very awry about it or the person has an aberrant personality.

-- B. Lewin, personal communication, 1982

This is the way Dr. Benjamin Lewin, editor of the respected journal *Cell*, explains how a senior scientist can be unaware of research irregularities in his own laboratory. Fortunately, deliberate falsification of research data is exceptional; for fraud is anathema to scientific endeavour. In every aspect of their work, scientists rely heavily on the techniques, methodology, knowledge, and theories developed by their colleagues. If they hope to advance their field, they must build on previous work – and must assume that their predecessors did the experiments they claimed to have done, and reported their results accurately.

Although this assumption has not been seriously disputed, there have always been cases of scientific deception. Analysis of these occasional and reprehensible acts of fabricating research evidence reveals several basic flaws in the way academic research is organized. Current and traditional scientific practices don't often prevent, and can sometimes even encourage tampering with data for personal gain. As long ago as 1830, Charles Babbage defined three kinds of fraud in scientific research: forging, cooking, and trimming.[1] Over the years, examples of each kind of malpractice have been exposed in many scientific disciplines,[2] and several recent cases of scientific finagling have become media sensations. Whether the proportion of dishonest scientists has increased, we can't know; but the public is certainly becoming much more aware of those who do tamper with their results.

A case of alleged deception involved a prolific post-doctoral researcher in the laboratory of Dr. Louis Siminovitch, at the University of Toronto in Canada. The post-doctoral fellow, Dr. Demetrios Spandidos, was working in the exciting new field of gene transfer. Simply put, his experiments were designed to transfer one piece of genetic material, DNA, from a certain type of cancer cell to a recipient, non-cancer cell, to determine whether cancer has a simple genetic basis. (This gene-transfer methodology, now called "DNA transfection", has since been perfected and is used routinely to study genetic characteristics.) Ultimately, the implications of this technique are enormous. Eventually, defective genes might be replaced by fit ones and malignant cells might be rendered benign by genetic manipulation.

This field was just breaking in 1977, when Spandidos reported that he had developed a new technique for transferring genes from one strain of animal cells to another. His new method of gene transfer seemed more successful than any other process.

At the Brookhaven Symposium on Biology in 1977, O.W. McBride and R.S. Athwal, two well-known cancer researchers at the National Institutes of Health in the United States, welcomed Spandidos and Siminovitch's results, and described the dramatic impact their work might have on the young field: "Spandidos and Siminovitch ... have made significant modifications in the procedure for chromosome uptake and concurrently greatly increased the frequencies for CMGT [Chromosome Mediated Gene-Transfer] (by about 100-fold).... This method will become extremely important if it can provide similar transfer frequencies in interspecies transfer systems."[3] It was, therefore, especially disappointing when Louis Siminovitch alleged that Demetrios Spandidos had never done some of their reported work. To understand how this might have happened without the knowledge of the senior scientist, it is important to remember the tremendous excitement and anticipation that surrounded this research in the mid-1970s. Spandidos was one of the pioneers of chromosome-mediated gene transfer, working at the leading edge of a brand-new field – the manipulation of purified genetic material.

Spandidos came to Louis Siminovitch's laboratory highly recommended by a respected scientist in the field who was a very good friend of Siminovitch's. And he worked very hard: "seven days a week, nine to twelve".[4] Before long, he was churning out volumes of data. In only a year and a half, he had produced six papers, which were co-authored with Siminovitch and published in top journals.[5] His impressive record prompted Siminovitch to recommend him for the Canadian Medical Research Council's prestigious Centennial Fellowship, which he received.

Spandidos continued to be enormously prolific, but Siminovitch began to wonder what all this productivity was based on: "I had a technician working with him and she felt he just didn't know how to do experiments. And that got me a little nervous because you would have to have real golden fingers to do as much as he had done so accurately and so well...."[6] Nevertheless, Spandidos continued to produce the data and the publications to add to his and Siminovitch's growing list.

Then, in early 1978, a post-doctoral researcher, Bill Lewis, spent several months in Siminovitch's lab trying to apply Spandidos's techniques exactly as they had been reported. Despite repeated efforts, he failed to transfer genes from one batch of cells to another. Finally, he worked with Spandidos on one experiment. While Spandidos claimed it was a success, Lewis charged that sloppy technique had allowed the genes in fragments of the original cancer cells to contaminate the culture containing the recipient cells, i.e., that the supposedly transferred genes had never been actually incorporated into the recipient cells.[7] (Subsequent efforts to repeat Spandidos's specific experiments on the part of Siminovitch and Lewis also failed,[8] but the general gene transfer technology has proved successful.)

Another incident further eroded Siminovitch's confidence in Spandidos, and finally resulted in Spandidos's expulsion from the laboratory:

> One experiment just required an enormous amount of material, and I checked for the amount of material he was using and it wasn't there. And then the referee's report came and it said, "I'm amazed at how much this man has done, because it would have taken this and this many plates," et cetera. That was the [last] straw. So I called him in and ... he says, "I haven't made anything up," and I said, "Well you would have needed six hundred plates for this experiment" – I don't remember the exact number and I didn't say 600 plates – he says, "Well they were there," and I said, "Well I looked at the thing and they weren't there," and he says, "Well they were there," and the next day he came in to see me and he says, "It wasn't 600 plates because I used bigger plates, it was 300 plates," so I said, "Well, I didn't see 300 plates," he said, "Well, I often used bottles too." And [he

continued with] that sort of rationalization....

I told him I didn't want him to do any more experiments but I gave him two or three months to clean up his affairs....

What no one's ever really understood is that I never let him go because we couldn't repeat the experiments. I let him go because he wasn't doing some of the experiments. I never did know how much was true and how much wasn't true.[9]

In June 1978, Demetrios Spandidos returned to Greece, claiming that "pressure from Siminovitch" (perhaps through unofficial conversations with colleagues) had made it impossible for him to secure another position in North America;[10] but he continued to plead his case. He sent out a form letter to the Medical Research Council, as well as to the editors of several journals in the field. In it, he asserted his innocence and asked for their support in gaining "the fundamental right to defend myself in front of an independent scientific committee to present in more detail my side of this extremely important issue".[11] Spandidos never got that opportunity.

The lack of official recognition of these allegations ultimately allowed Spandidos's career to prosper. At first Spandidos was forced to leave Canada, and was deprived of his MRC scholarship as a result. He lost the glowing job prospects that Siminovitch's influence and praise might have secured him. Likely job offers at renowned American biomedical research establishments failed to materialize after Siminovitch asked Spandidos to leave his laboratory. Yet less than a year later, Spandidos's MRC fellowship was restored to him when he secured a position in the laboratory of Dr. John Paul, at the Beatson Institute for Cancer Research in Glasgow. He has published prolifically since then, authoring seven articles in 1984 alone, one of them in the prestigious journal *Nature*.[12]

Unfortunately, the Spandidos-Siminovitch case is not an isolated example of research misconduct. A rash of scandals has erupted recently in prestigious biomedical research laboratories in the United States and Canada.[13] Within the past several months alone a team of scientists who had published a major research article in *Science* admitted that "those biological data ... are incorrect, and we wish, therefore, to retract the data and the conclusions based on them":[14] the U.S. National Institutes of Health (NIH) severely censured Dr. Charles J. Glueck, a medical scientist they had funded, whose published work they concluded incorporated "extensive inconsistencies and errors ... [which] represent substantial departures from generally accepted standards of scholarship";[15] and a panel of the U.S. National Institute of Mental Health found that Dr. Stephen E. Breuning, a well-known researcher of mental retardation, "knowingly, willfully, and repeatedly engaged in misleading and deceptive practices in reporting results of research".[16] (Several institutions changed the treatment of patients to conform to Dr. Breuning's results, which have now been judged to be fraudulent.)

Between 1980 and 1982, the U.S. National Institutes of Health was notified of forty-five allegations of research misconduct, and in 1983 similar charges were made to them a couple of times each month.[17] In 1976, *New Scientist* surveyed its readers, who are, predominantly, engaged in scientific work, on the subject of cheating in science. More than 50 per cent of the respondents reported personal knowledge of intentional bias in research; a further 17 per cent learned of it from colleagues who had direct contact.[18] Dr. Robert G. Petersdorf, Vice Chancellor for Health Sciences at the University of California at San Diego, blames the "highly competitive pressures in modern science". It is, he says, "similar to cheating on exams, where we keep finding scandals wherever we look for them".[19]

Despite Daniel E. Koshland Jr.'s confident proclamation in an editorial in *Science* in January 1987 that "99.9999 percent of reports are accurate and truthful",[20] there has been no comprehensive survey establishing what proportion of published scientific reports are either honest or dishonest.

In one of the most shocking of these cases, Mark Spector, a brilliant graduate student at Cornell University, fabricated a whole series of experiments.[21] Spector had worked in the laboratory of the eminent biochemist Efraim Racker, who, like Siminovitch, was a senior investigator heading a large research team; as well, he had many other scientific and administrative responsibilities. During the year and a half Spector worked at Cornell, his colleagues were impressed by his intelligence, technical expertise, and devotion to his work.

Spector's results were spectacular – they promised a unifying framework for the cause of cancer. They were presented in *Science*, under the ambitious title, "Warburg Effect Revisited: Merger of Biochemistry and Molecular Biology". The article began with a quote from G.K. Chesterton: "There are no rules of architecture for a castle in the clouds."[22] Unfortunately, Spector's promised panacea turned out to be just such a castle. His work was deception on a massive scale. After a colleague discovered in some of Spector's research material a radioactive isotope that had no business being there, Dr. Racker retracted their co-authored papers.[23] He worked long days in the laboratory, repeating all of Spector's work, to find out whether any of it could be reliably reproduced. Spector left the laboratory and withdrew from Cornell in mid-1981.[24]

Nor is the problem confined to North America. The eminent Swiss biologist Dr. Karl Illmensee of the University of Geneva was accused by his own lab colleagues of not adhering to standard protocol in his experimental work. (American researchers have also had difficulty repeating his spectacular success in manipulating mouse embryos.) Illmensee claimed to have cloned mice, to have incorporated human or rat genetic material into mice, and to have reared mice having genes from only their mothers. After one of his collaborators publicly discredited some of Illmensee's results in 1983, several commissions were set up to investigate Illmensee and his alleged achievements. The University of Geneva's external review committee drew no formal conclusion regarding Illmensee's motivation. They did, however, find that "the numerous errors and disrepancies found in the experimental protocols ... throw grave doubts on the scientific validity of the conclusions". The committee appointed by the Jackson Laboratories at Bar Harbor,

Maine, where Illmensee did some of his work in conjunction with Dr. Peter Hoppe, also found no evidence of fraud, but advised Illmensee and Hoppe to repeat their experiments using a more conclusive experimental design.[25]

These incidents raise a number of crucial questions, concerning their causes as well as their consequences. What is the long-term impact of alleged research fraud on progress in the field? What is the ultimate effect of such scandals on the careers of those involved? How can the scientific community find out what really happened in such cases, and what can be done to ensure a just outcome? Most important, what can be done to prevent similar incidents from occurring in the future?

The advancement of any field of science can be traced by the publications in it. While there is often a great deal of lively debate among specialists at conferences and seminars, science is now too international and vast a venture to be communicated primarily by word of mouth: scientists must rely on publications to keep up to date. Consequently, the possibility that published work is ill-founded can have widespread repercussions.

There are few established rules for dealing with the allegation that some published papers are based on incorrect or inadequate data. Dr. Benjamin Lewin, editor of *Cell*, believes that a formal retraction is the wisest course:

> If we learn that there is some degree of doubt about some series of experiments which we have published, we usually say to the senior author that it would be proper to publish some sort of statement in the journal to clarify the position.... there's a responsibility to a much wider range of people who read the article, who don't keep abreast of everything in the field and who can be easily misled – and there's a responsibility to set the record straight. We do encourage authors in those circumstances to publish a retraction.[26]

This is exactly what Efraim Racker did. He retracted his co-authored paper with Spector as soon as he suspected any wrongdoing, even before he knew how much of their published work might stand up to his attempts to repeat it. In a formal letter of retraction to *Science*, where one of the suspect articles had been printed, he declared:

> ...an article was recently published in *Science* (17 July, p. 303) that I coauthored with Mark Spector, a graduate student in my laboratory. I feel compelled to withdraw some of the claims that we made in that article.... We are now checking all published data, and it will take us many months before we know what is correct. We suspect that some of the data dealing with cells transformed by various tumor viruses are incorrect. We did not deal with these experiments in the *Science* article, but they were subjects of a paper in press which we are withdrawing. They were also presented by me and others in seminars, and I wish to withdraw these claims until we can verify them. [27]

He sent a similar letter to *Cell*, which had also published a paper based on Spector's work.[28]

Louis Siminovitch adopted a strikingly different course. In the two years that Demetrios Spandidos had worked in Siminovitch's laboratory, they had co-authored five papers in such journals as *Cell*, *The Proceedings of the National Academy of Sciences of the U.S.A.*, *Brookhaven Symposium on Biology*, and *Nature*.[29] Two more articles had been submitted to journals and two were in press in April 1978 when Demetrios Spandidos was asked to leave Siminovitch's laboratory.[30]

These four papers were withdrawn prior to publication, but the published articles were not formally retracted in the usual way: a letter of retraction in the journal of initial publication. Instead, Dr. Siminovitch initially relied on word of mouth: he phoned several of his friends to let them know what had happened; and in a public lecture in the summer of 1978, he voiced his suspicions regarding the gene-transfer research.[31] Two years later, in his next publication on the subject, he included the following disclaimer: "One of us (L.S) was involved as a co-author in a series of papers in which the transfer of a variety of other genes by means of metaphase chromosomes was described [see note 5, Spandidos and Siminovitch 1977a, b, c, d, 1978a]. We have not been able to repeat those experiments, and we have reason to believe that the technologies described in those papers do not result in successful gene transfer."[32]

Somatic Cell Genetics, which published the article quoted above, was not one of the journals that published the six suspect articles. Dr. Siminovitch gave two reasons for his chosen course of action: "I just didn't know how much was true and how much wasn't true: where do I stop and where do I begin, saying that they're wrong?... But the second reason of course is that I was concerned that if I said that, I would have a lawsuit.... How do you prove that you hadn't done an experiment? I mean, he could always say that the conditions have changed and the salts have changed, and you could go on for years that way."[33] One thing is certain: if the scientific community at large is informed of the concern promptly and effectively, the damage will be minimized. But this happens all too rarely. In the case of Spandidos and Siminovitch, such action was not taken and, incredibly, the several articles co-authored by Spandidos and Siminovitch are still cited in publications on related matters.[34] In 1980, for example, they were referenced very favourably in a widely read review of chromosome-mediated gene transfer.[35] If Dr. Siminovitch's suspicion is correct, that some of the experiments were not performed, many researchers may be on expensive and time-consuming wild goose chases, trying to follow up on Spandidos's apparently exciting results.

The procedures followed in several similar scandals in the United States have ensured that most allegations of scientific fraud in that country are eventually investigated and judged, and that some disciplinary action is taken when cheating has been established. Thorough audits by external review teams have determined the extent of the deception in various recent American incidents. The U.S. National Institutes of Health, a major research funding agency, has been particularly rigorous in acting on allegations of research misconduct. They have investigated and publicly reported on a long series of cases of research malpractices, most recently, in May 1987, by excluding Dr. Charles Glueck from NIH funding and peer-review service and by notifying the publishers of his discredited work.[36] The National Institute of Mental Health has been even more aggressive in its review of Dr. Stephen Breuning's work by distributing its report to his employer and to professional boards and associations and by initiating action to recover grant funds he had misused.[37] Such action has not always been immediate, however; investigations generally begin several months after the concerns have been raised.

Almost a whole year elapsed in the case of Vijay Soman, an assistant professor at Yale Medical School who was accused of plagiarizing a manuscript he had reviewed. Despite the assurances of Soman's supervisor and then vice-chairman of the Department of Medicine, Dr. Philip Felig, that Soman's data were credible, a professor at Harvard Medical School, Dr. Flier, was called in to thoroughly audit the data that had been published in an article Soman and Felig had co-authored. This article was found to misrepresent the experimental results; Soman was asked to resign immediately, which he did. His supervisor, Felig, was forced to give up his recently assumed and very prestigious post as chairman of medicine at the Columbia College of Physicians and Surgeons in New York. All the other raw data that Soman had produced at Yale were subsequently audited and found to be seriously inadequate.

Eventually, eleven papers based on these experiments were retracted.[38] (The pitfalls of the publication process are discussed further in Chapter Six.)

A similar procedure was followed in the case of John Roland Darsee, who was caught concocting data in May 1981 at Harvard Medical School's Cardiac Research Laboratory.[39] When it was learned that Darsee's work on a major National Institutes of Health (NIH) study was also suspect, the university set up a committee of senior academics to investigate the affair. The NIH also appointed four university-based cardiologists to do an independent audit. As a result of these investigations, Darsee was barred from the National Institutes of Health and disqualified from funding for ten years.[40]

Sadly, Canadian official behaviour is in stark contrast to the relatively vigorous U.S. investigations of allegations and subsequent disciplinary actions. Spandidos and Siminovitch seem to have been protected by the lack of any conclusive investigation of the suspicions concerning their co-authored work. Spandidos has always asserted that the work he did in Siminovitch's lab was valid, and has since cited it favourably. When he was asked to leave the laboratory by Siminovitch, one of the first things Spandidos did was write to the Medical Research Council of Canada (MRC), pleading his innocence and requesting that the council set up a committee to investigate the allegations against him. Dr. James M. Roxburgh, then director of the grants program at the MRC, which had paid for the questionable research effort, replied to Spandidos, refusing his request. Roxburgh later defended this action, saying: "It was just not considered necessary or desirable to set up such a committee.... These allegations were simply hearsay to us. They were reported to us by Spandidos. They were never made publicly by his supervisor or to us formally by his supervisor."[41] Dr. Roxburgh did admit that, although there had been no formal allegations made by Louis Siminovitch against Demetrios Spandidos, he had "known Lou Siminovitch for many many years, and naturally ... there were informal conversations off the record".[42] The MRC feels that it's up to the supervisor to establish the truth in these cases. But Louis Siminovitch was clearly reluctant to carry out his own investigation, partly because "it might have taken even more than two years",[43] and partly because it may, in the end, have been inconclusive.

Many scientists argue that the consequences of incidents such as the one involving Spandidos and Siminovitch are ultimately insignificant. Dr. Robert Weinberg, an associate professor at the Center for Cancer Research at MIT and a pioneer in the development of gene-transfer techniques, doesn't view the Spandidos and Spector affairs as a "disaster for science....

They're amusing, and they have an interesting human element to them, but in a field in which there are many scientists interested, these two incidents represent small tempests in a teapot and are soon glossed over. One soon finds out what the real answers are and forgets these minor impediments which happened along the way."[44]

Other scientists try to rationalize such controversies, assuring themselves and others that actual perpetrators are incredibly rare, pathologically abnormal individuals. No sane scientist, they say, would risk the inevitable censure of the academic community by engaging in such a risky venture. The consequences of Soman's large-scale distortion of data are undoubtedly grave. But for a young scientist working largely independently, with the prospect of a glittering career, the risks may seem meagre when compared with the potential rewards.

Strong similarities between the cases that continue to come to light[45] point to a problem inherent in the way scientific research is organized. The system of academic-career advancement, coupled with the laboratory hierarchies prevalent in most research establishments, creates both the motive and the mechanism for scientific malpractice. A detailed analysis of these structures is essential to understanding how to prevent future scandals of this kind.

Many of these cases have occurred in exciting new areas of biomedical research, where the potential achievements are enormous; so too is the pressure to be the first to prove its initial promise. Competition for the prestige, positions, and generous salaries awaiting successful scientists in these fields is, therefore, very stiff. After Vijay Soman was caught having doctored his data, he was reported in *Science* to have blamed his transgressions on "the cut-throat rate of research". "My actions," he was quoted as saying, "were done in the midst of significant pressure to publish these data as fast as possible so as to obtain priority."[46] Indeed, competition to enter and secure a position in academic science has been increasing; across-the-board demand for this kind of employment has remained high, despite cuts to research-funding agencies. This competitive atmosphere is heightened for those young scientists working in other scientists' laboratories who are seeking their crucial first positions as independent researchers.

Dr. Louis Siminovitch admits that this can pose difficulties for many young researchers:

> Post-doc years are probably the best years of a scientist's life, or should be. They're the most traumatic for them, because they've always got to think, there's a job problem at the end.... As a post-doc you're given money, you're given space, you're given all of biology there to be discovered, and you're young and what else? ...essentially it is an idyllic situation. Except for the problem that at the end of the road you've got a job to find, and of course, you're deciding whether you're going to make it. Not everybody responds to ... that situation.[47]

It is no mere coincidence that all the cases I have referred to involved young researchers, in their late twenties or early thirties, on the brink of emerging from the laboratories of eminent senior scientists. While studying with such people certainly enhanced the young researchers' credibility, it also meant they were forced to work much more independently than they would have in the more modest labs of less eminent scientists. For as they become better known, senior scientists usually spend less and less time in the laboratory milieu.

A successful scientist's growing reputation is almost always accompanied by mounting administrative commitments. To provide a large laboratory with the necessary space, equipment, and financing demands time: time to apply for grants, to hire staff, to screen new students, and to maintain a favourable position within the institution. The director of a large research laboratory also has responsibilities: to the university department, to the agencies providing funding, and to the community at large. Senior scientists participate in academic and professional committees to keep abreast of the latest developments in the field and to maintain the political influence that can ease their and those of their students' future careers.

Grant-awarding bodies and scientific journals need well-established scientists for their peer-review committees, as they naturally want the best advice on what work to fund or publish. Well-known scientists who maintain some research involvement are also sought after on political or educational advisory committees and on journal editorial boards. The external commitments of established research scientists mount with their reputations and, inevitably, these investigators spend progressively less time doing hands-on research. This trend is supported by an investigation of the science faculty of U.S. graduate schools, who are generally relatively well-established in their fields: the National Science Foundation in the United States reported that "scientists and engineers on the faculty of graduate schools in the United States spend *an average* of only 16 hours a week on research".[48]

Louis Siminovitch is a good example of the successful senior scientist. In 1981 he was awarded the $25,000 Gairdner Foundation Wrightman Award for outstanding leadership in medicine and medical science in Canada. The citation included the following statements: "His awareness of the impact of science on society is as important a part of his scientific makeup as is his enthusiasm for new ideas and his insistence on high standards in his experimental work."[49] In the same year, he was also one of the first recipients of the newly established Izaak Walton Killam Memorial prizes, in recognition of his "distinguished lifetime achievement and an outstanding contribution to the advancement of knowledge".[50]

His list of external advisory positions is impressive. In 1982, he was vice-president of the National Cancer Institute of Canada (NCIC); he was on the executive of the Medical Research Council of Canada; on the Ontario Council of Health; on the executive of the Ontario Council of Health; on the board of the Ontario Cancer Treatment Research Foundation; on the Advisory Board for Cancer Treatment and Prevention of the National Cancer Institute in the United States; chair of the Canadian Broadcasting Corporation Advisory Committee on Science and Technology;

chair of the special Terry Fox Committee of the NCIC; an editor of about seven journals; on the board of the Mount Sinai Research Foundation; and on the advisory board of the Connaught Medical Research Laboratories. He subsequently became director of the new research unit of Toronto's Mount Sinai Hospital.

As Siminovitch himself admits, "I've got my fingers in quite a few pies."[51] This administrative work doesn't itself produce an iota of data. What it does is ensure that Siminovitch has the power to choose the direction of research, the supply of money and equipment required to produce the results, and the access to publications that he needs to perpetuate his laboratory empire.

Siminovitch's preoccupation with professional and administrative activities may well have prevented him from having a very detailed knowledge of the work in progress in his lab: "So in fact it's very common, all over science, that the professors don't work in the lab, or very little, and that the work is done mostly by students, post-docs and technicians.... I'm a bit unusual in the sense that I have quite a bit of extraneous activity, compared to most people. But you know I work pretty hard, I'm here all the time, when I'm not out of town, at a meeting or something like that, but I don't think I'm overly unusual."[52] The political nature of senior scientists' work brings with it definite benefits. They maintain the knowledge and the friendships to get their own and their associates' work published and funded. One of their attractions for young scientists is that their connections with journal editors can greatly ease the often lengthy and frustrating business of getting research reports published (how greatly, is discussed in detail in Chapter Six). Dr. Lewin, editor of *Cell*, describes the trust and high expectations which await papers co-authored by established scientists:

> Look at it this way. You work in some area of molecular biology. You publish very distinguished papers for a series of years, all your papers are well received, you make what are acknowledged to be significant contributions to the field. When you write another paper, that paper has a certain degree

> of credibility. And that's undeniable. By virtue of
> the reputation of your laboratory in the past, when
> you write a new paper it brings to that an enhanced
> degree of credibility. Perhaps an enhanced degree
> of credibility is too strong – it brings to it a belief
> that what you say you have done in this paper is
> exactly what you *have* done.[53]

In a kind of Catch-22 situation, this borrowed or vicarious
credibility may allow the work of a relatively junior scientist to escape serious critical scrutiny altogether: the senior scientist
co-authoring resulting papers may not have the time or inclination
to go over the research procedures, analysis, and interpretations
in detail, and journal reviewers may also neglect to investigate
these matters as intensively as they should, assuming the
respected senior scientist has already done so. This seems to have
been the case for many of the 109 papers published by Dr. John
Darsee and his 47 co-authors. A detailed analysis of these papers,
by Walter Stewart and Ned Feder at the National Institutes of
Health, published in *Nature* in 1987, found that all but two of the
papers contained from one to thirty-nine errors and discrepancies
per paper (an average of about twelve).[54] All of these "lapses from
generally accepted standards of research",[55] which included unacknowledged republication, misleading statements,
discrepancies in figures, and impossible data (e.g., parents aged
eight or nine), were presumably missed by co-authors, journal
reviewers, and editors alike. Dr. Darsee, now judged to have
repeatedly published fraudulent work, may have been responsible
for many of these "lapses", but the presence of reputable co-authors
likely meant that these papers did not receive a very critical recep-
tion.

There is a kind of unwritten contract between the junior and
senior scientist: the senior scientist secures the research funds,
using his or her reputation to ensure generous grants to cover
the cost of equipment, chemicals, and even the junior
researcher's salary. In exchange, the junior investigators churn
out data and draft the papers, which they and the senior
scientists co-author, to provide the senior researcher with proof

that his or her research funds are well spent (see Figure 1).

A hierarchical lab structure is established: the junior researchers work in the laboratory, largely unsupervised, and merely meet with the senior scientist to discuss experimental results and revisions to draft papers. It is easy to see how intense competition for rich scientific rewards in a hierarchical laboratory setting could conspire to tempt ambitious young scientists to concoct especially interesting data. These serious flaws in the structure of current scientific research make an internal checking system for science all the more necessary.

Scientists place great faith in what is called "experimental replication": if research work cannot be reproduced exactly by an independent investigator, it becomes suspect. But this ultimate

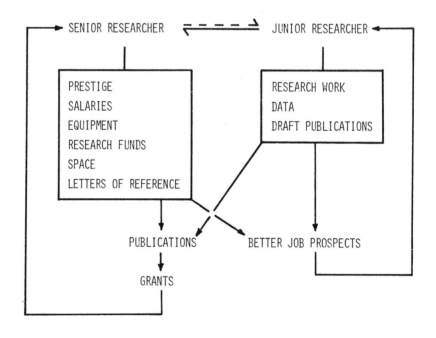

Figure 1

arbiter of scientific "truth" cannot always be relied upon. If the experiments are tedious to perform and their results not surprising, they are likely to be assumed to be correct without follow-up. It's clearly more exciting and prestigious to author original work than merely to report on the reliability of someone else's data, and for this reason, exact replications of experiments are not popular. In an analysis of publication decisions, T. Sterling, professor of computer science at Simon Fraser University in Vancouver, noted that "significant results published ... are seldom verified by independent replication".[56] This means that fraud not discovered in the perpetrator's laboratory may remain undetected for years, and scientists may be building on false results. As Sterling notes: "The possibility thus arises that the literature ... consists in substantial part of false conclusions."[57]

Of course, it is impossible to know how often data are fabricated and whether the cases of proven fraud are the tip of an iceberg of scientific subterfuge or merely rare, atypical incidents. Nevertheless, the similar circumstances of these cases indicate that there may be something in the lab situation that spawns them. The combination of pressure, competition, large potential gains, and a lack of close supervision may convince some junior scientists that the risk of getting caught is minimal compared with the possible rewards of deception. All research institutions should, therefore, develop consistent and enforceable procedures for preventing and dealing with fraud. Several American institutions have established mechanisms for ascertaining the validity of questionable results and for censuring guilty parties when detected.[58] (For projects funded by the U.S. National Institutes of Health, such procedures are now required by law.)[59]

Dr. Eugene Braunwald, who worked with Dr. John Darsee, later found to have engaged in fraudulent research practices, goes so far as to urge investigation and criminal prosecution in alleged cases of dishonest research reporting.[60] At the very least, clear, comprehensive laboratory protocols, requiring, for example, open data books and the involvement of the laboratory directors in experimental design and analysis, would help to define appropriate behaviour and identify individual responsibility.[61] But independent institutional initiatives will not change the system of scientific advancement and rewards that gives cheating its appeal; they can only make it more difficult. Therefore what is much more important is to develop a style of science that provides less incentive for deception. For this to be effective, universities and other research institutions must collaborate in a concerted effort to make researchers more accountable, labs more egalitarian, and to make the rewards of scientific success, publications and grants more equitably available to all those conducting scientific enquiries. Research dollars will always be scarce, relative to the demand for spending them, and some open, fair, and responsive procedures should ensure that they go to those scientists with the best research ideas. The crucial issues of how research merit is demonstrated in publications and then judged by granting committees are the subject of the next two chapters. The reforms recommended in them would go a long way towards undermining the lab hierarchies that can make the academic hustle an unsavoury exercise.

Chapter 6
GETTING INTO PRINT

"Publish or perish" – the notorious academic cliché embodies the cynicism, even the resentment, of scientists faced with the constant need to show concrete evidence that research time and financial support have not been wasted.

Of course, research discoveries can help no one if they are not disseminated. But the drive to publish as many papers as possible regardless of how much they contribute to the advancement of the field or the public at large has become a scientific obsession. Indeed, it can be argued that publications are the most important requirement for scientific success. Postdoctoral positions, job offers, promotions, and the acquisition of research funding and talented assistants all depend on a robust publication record. Canadian granting committees traditionally expect one or two articles per year by individuals (and at least double that for heads of laboratories) from efficient and productive researchers.[1]

This emphasis on publication, whereby scientists are evaluated by the sheer numbers of their articles or books, has led to the proliferation of scientific journals and to intense competition to publish in the most reputable of them. In 1986,

The New England Journal of Medicine, a prominent biomedical publication, had a rejection rate of about 88 per cent. Each year they receive approximately one hundred more manuscripts than in the preceding year.[2] Robert Campbell, of Blackwell Scientific Publications in Oxford, reports that the most sought-after sociology journal rejects 90 per cent of all submissions, although 60 per cent of them do find their way into print elsewhere within five years.[3] There is some variation in rejection rates among journals in different disciplines: highly regarded psychology publications typically have an 80 per cent rejection rate;[4] a comparable physics journal, *Physical Review Letters,* rejects 55 per cent of the manuscripts it reviews.[5] Nevertheless, publication in the most sought-after journals is never easy.

The desperate need to publish, combined with the difficulties of doing so, have resulted in a number of rather unsavoury practices. Scientists routinely subdivide one piece of research work into several small chunks to be published separately. (Some journals actively encourage this, by limiting the length of articles they will publish).[6] While this practice certainly inflates a scientist's list of publications, it does little to help other researchers; they must leaf through several journals to assemble information on a single research project.

As the length of papers shrinks to what has been dubbed by science commentators the "least publishable unit"[7] the list of authors grows. In order to get maximum professional mileage from research, the number of co-authored papers (two or more collaborators sharing credit for joint work) has skyrocketed.[8] Laboratory heads or departmental chairmen may get automatic co-authorship of papers published by their staff, even if their personal involvement was negligible. This problem was highlighted by a recent panel of the U.S. National Institute of Mental Health, which concluded in May 1987 that Dr. Stephen E. Breuning had committed a scientific fraud.[9] The panel found that Breuning's co-authors sometimes had "little or no involvement in the work reported", and urged editors to insist that authors "bear responsibility for the integrity of [their] published work".[10] Walter W. Stewart and Ned Feder, in a recent study of the fraudulent papers published by Dr. John Darsee, found that thirteen out of a sample of eighteen papers they scrutinized had what they dubbed "honorary" authors who had little direct involvement in the reported research.[11] Interestingly, these "honorary" authors co-authored more papers than non-honorary co-authors, which would, of course, serve to enhance their already well-established careers and thus remove them still further from day-to-day research work.

In some cases, the pressure to provide visible proof of productivity has even led to multiple publications of the same or very similar material.[12] As Carl J. Sindermann, a successful marine biologist, writes in his light-hearted but cynical book, *Winning the Games Scientists Play:* "Administrators are impressed by numbers and weight, so rewrites of the same material under slightly different titles are in order, as are impressive covers on reprints and the citation of abstracts, notes, and book reviews in an annual list of publications."[13] All these techniques give meagre substance the *appearance* of prolific, respectable output.

One recent example of this occurred when two very similar papers, by Claus Christiansen and Bente Riis, of the University of Copenhagen, were published in January 1987 in *The New England Journal of Medicine* and *The American Journal of Obstetrics and Gynecology*, neither of which referred to the other article. When readers wrote to the two journals informing them of this problem the editors of *The New England Journal* published the letters from readers, a response from the authors, and an editorial statement condemning republication of material, publicly reprimanding the authors, and warning them that further infractions would lead to denial of publication of their future work in the journal.[14] Another case involved J.M. Lauweryns, J. Baert, and W. De Loecker, from the laboratory of histopathology and the laboratory of pathological biochemistry at the Catholic University of Leuven Medical School in Belgium. They published strikingly similar papers in the *Journal of Cell Biology* and *Cell and Tissue Research* in successive years with no reference in either paper to the existence of the other.[15] The reported number of times each experiment was performed or replicated varied slightly in each report, but the

Journal of Cell Biology paper was really just an abbreviated version of the other report. On another occasion one of the authors, J.M. Lauweryns, seems to have reported some other experiments twice, repeating figures and passages of text in articles published with co-author M. Cokelaere.[16] On two occasions[17] K. Izutsu published identical figures. Another author, H. Sato, republished segments of his own text.[18] In all these cases there was no indication that the items in question were already published.

Duplicate publications have become so common that several journals have taken steps against authors of papers submitted to more than one journal.[19] *Cell* and *The Proceedings of the National Academy of Sciences* have agreed not to publish any articles for three years by any author who publishes the same material in both journals; *Science* editor Philip Abelson wrote an exceptionally stern editorial condemning the practice:

> The drive to expand bibliographies is understandable, even if some of the methods are not forgivable. Competition for research funds is strenuous. A good publication record is important. Today, in fast-moving fields, studies are often conducted by large predoctoral and postdoctoral teams. Most of the members' names on a publication may wind up cited in the "*et al.*'s." If there are multiple publications, each member can have a turn at being first author. Another incentive for multiple submissions is the regrettable delay that occurs when a journal is slow in processing manuscripts. There is a temptation to submit papers to several journals with the plan of withdrawing from some after acceptance by one. But excessive submissions have deleterious effects.... In cases of duplicate publication of original research findings involving *Science* and other major periodicals, we will, after careful examination of the facts, consider joining in concerted action against the offender.
>
> We hope that punitive actions will not be necessary. The prospect of them is disagreeable.

> But the integrity of scientific publication must be
> maintained, and offenders must be made aware
> that they have more to lose than to gain by their
> behaviour.[20]

This threat of retroactive sanctions constitutes an admission by editors that they will likely fail to recognize duplicate manuscripts before publication.

The New England Journal of Medicine has adopted an even more extreme position. The editor, Arnold Relman, is especially concerned about what he calls the irresponsible dissemination of unconfirmed medical claims by the press.[21] In an effort to discourage early popular reports on new breakthroughs, the journal will refuse to publish any work known to have been reported elsewhere, either in the popular press or in academic literature. Ultimately these measures could not only delay publication of important new information but also restrict knowledge to a specialized audience. The implications are ominous. Another alternative, to submit a manuscript to several journals and subsequently withdraw it from all but one, would result in speedier publication. The possibility of protracted, expensive legal battles makes this option less attractive, however, as contractual obligations might be required to prevent multiple publications.

Republication by scientists of their own work without reference to their source is clearly reprehensible, but publication of someone else's work without credit is much worse. Plagiarism is a serious academic offence,[22] and, regrettably, one that may be widespread. Daryl Chubin reports that 20 per cent of cancer researchers who had been refused funding by the U.S. National Institutes of Health had experienced pirating of their data or ideas for use in reviewers' subsequent proposals.[23] Elias A.K. Alsabti managed to forge an entire academic career, and a highly successful one, using fabricated credentials and other people's work. *Betrayers of the Truth*, by William Broad and Nicholas Wade, describes how Alsabti worked on research fellowships in such prestigious U.S. labs as the M.D. Anderson Hospital in Houston and Temple University in Philadelphia, publishing more than sixty papers between 1977 and 1979 as his own work. Many were lifted verbatim from other people's grant applications and from foreign journals. One of the researchers who worked with him explains how Alsabti got away with the plagiarisms for so long: "Alsabti knows the system, and he knows nobody wants to get dirty. Nobody wants to be the first to say, 'Hey, this guy's a fake.'"[24]

A more spectacular case of plagiarism involved a European pensioner who copied more than one thousand scholarly articles and published them in other journals under his own name. He even managed to republish several articles as his own in the very journals that had originally printed them![25] (The exploits of Vijay Soman, at the Yale University Medical School, have been described in Chapter Five.)

Of course, there is no definitive proof that plagiarism is on the increase, but current pressures to boast impressive curricula vitae must make it increasingly tempting. None the less, some argue that the problem will solve itself, if we clamp down on the rare offenders. This is the "bad apple in every bushel" argument: just chuck out the rotten one and, supposedly, the rest will be unaffected.[26] If only it were that simple. Corruption doesn't spring from nowhere: both temptation and opportunity must present themselves. Ample temptation is provided by the unprecedented rewards of scientific recognition. The opportunity for deceptive publication practices lies in the inordinate workloads of manuscript reviewers and editors; all too often, they are unable to keep up with articles published in their field and therefore fail to recognize material that is not original.

All scientists have an honourable motive for promoting their own work. They do it not only to advance their careers, but to publicize the ideas, theories, and fields of study they hold most dear. The incentive to lend credibility to favoured hypotheses with numerous publications adds a further inducement for authors to recycle old results. Moreover, reviewers and editors can easily, if inadvertently, usher an offending manuscript into print. They are usually active researchers, and, inevitably, they view a newly submitted manuscript with minds weighed down by personal preferences, assumptions, and concerns. Scientists reviewing manuscripts that challenge their preconceptions may reject them out of hand, with little regard for their contribution to the field. On other occasions, swayed by sympathy for the authors or their conclusions, reviewers may not be as critical as they should.

There are signs that scientists are beginning to accept the concept that they might have vested interests in the ideas or information presented in manuscripts they are asked to review. The Committee of Editors of Biochemical Journals of the International Union of Biochemistry, for example, has formally adopted the position that "all manuscripts received in the editorial office should be considered privileged [confidential] communications, and be so identified".[27] Such a warning, and the sanctions imposed on authors who republish material, indicate that journal editors are aware of their own inadequacies. Clearly, editorial procedures for manuscript evaluation often fail to reveal authors' deception.[28]

Journal editors traditionally vet manuscripts submitted for publication using what is called "peer review".[29] Manuscripts are sent out for critical comment to several established scientists in the author's field. The editor then relies heavily on their judgement when deciding whether to accept a manuscript. Dr. Stephen Ceci, a psychologist at Cornell University who has studied peer review, explains the rationale for this process: "The idea is that these scientists are in the best position to assess the merit of the work since they're active investigators in that area themselves. So the two reviewers will receive in the mail copies of the manuscript. They will assess it on the basis of its 'scientific quality.' That's a broad term that includes things like how well it's designed, how appropriately it's analyzed in statistical terms, how clearly it's written, how it all hangs together, and so on."[30]

Normally, the authors' names are known to the reviewers, but the reviewers remain anonymous. Benjamin Lewin, editor of *Cell*, explains why:

> It's felt that reviewers will not feel free to make objective comments on a paper if their names are going to be revealed to the author. People are very reluctant to – how shall I put it? – sign their name to very critical comments. If the paper is great you have no problem saying, "This is a great paper. It should be published. I am Joe Smith." If you think it is not a good piece of work you're much more reluctant to say so – it creates animosity. There is always the risk that the author of the paper will retaliate some day. It brings a level of personality into the conduct of science that is felt shouldn't be there.[31]

For most existing scientific journals some form of this peer-review system is unavoidable. The alternative, to commercialize the entire process and operate it according to market and legal imperatives, would place a high financial premium on subscription and possibly publication as well, thus restricting access to authors and readers alike. Clearly, the demand to publish in widely read journals will always far outstrip the number of articles they can accept, and some method is needed to select which scientists and what research should gain access to the most desirable periodicals. Furthermore, it is probably safe to assume that most scientists are people of basic integrity, even though they may be swayed, usually unconsciously, by personal, professional, and political goals. Finally, the ultimate tests of time and reproducibility, while nothing like as speedy or effective as most scientists hope, will, in the end, challenge studies that are inconsistent with others in the field or which can only be accomplished by specific researchers.

In spite of these strengths, however, the current procedure for manuscript review has profound weaknesses. The reliability and fairness of the peer-review system was tested in a provocative experiment by Stephen Ceci and Douglas Peters, another American psychologist. At the time, both Peters and Ceci were based in the University of North Dakota. Once, while commiserating over their difficulties in getting their articles published, they speculated that their problems might arise not solely from their own inadequacies but also because both they and their university were relatively unknown. They decided to test this hypothesis: they chose twelve articles published by authors from prestigious institutions and resubmitted them to the reputable journals that had initially published them. Peters and Ceci concealed the identity of the authors, substituting fictitious authors and imaginary, low-status-sounding institutions. Only three of the thirty-eight reviewers – twelve editors and their chosen referees – detected the deception; of the nine papers that went on to be formally refereed, eight were rejected! In several cases the grounds for rejection were given as "serious methodological flaws".[32]

Many scientists have noted that Peters and Ceci's work was itself flawed: they used a small sample, only twelve papers, and they didn't balance their study by resubmitting the papers under the names of authors from ostensibly high-status institutions. None the less, Peters and Ceci have shown at the very least that reviewers from the same journal (presumably different reviewers on the two occasions the manuscripts were reviewed) rarely agree on the merits of a given paper.

A more recent study of this problem, reported in January 1987, supports the implication of Peters and Ceci's study, that reviewers are frequently negligent in carrying out their responsibilities. Two researchers at the National Institutes of Health reviewed 109 papers published by John Darsee, whose work has now been discredited as fraudulent. They found an average of twelve errors and discrepancies in each of these papers, many of them obvious inconsistencies that should have been noticed by journal reviewers or editors.[33]

Several other investigations confirm that reviewer agreement is rare. David Lazarus, editor-in-chief of the *American Physical Society*, asserts that "Even in my science, physics, which is by common consent (however misplaced!) regarded as far less

117

'subjective' than psychology, there is no way that we can run a journal with even far higher acceptance rates (45 per cent for *Physical Review Letters*) without encountering enormous discrepancies between the opinions of different referees. In only about 10-15 per cent of cases do two referees agree on acceptance or rejection the first time around – and this *with* the authors' and institutional identities known!"[34] Another team of researchers (Stephen Cole, Jonathan R. Cole, and Gary A. Simon) studied peer review by granting committees (the committees that decide whether a grant proposal should be funded), and they reached similar conclusions. (Chapter Seven addresses this topic in greater detail.) Publishing their findings in *Science*, they noted: "The degree of disagreement within the population of eligible reviewers is such that whether or not a proposal is funded depends in a large proportion of cases upon which reviewers happen to be selected for it."[35]

There is little doubt that the peer-review system has its problems.[36] Reviewers aren't always conscientious, and there is no guarantee that they are fully aware of the literature in their field. As a result, multiple publications of the same article and plagiarism often go undetected. Furthermore, reviewers disagree with one another. This is normal in any academic discipline – controversy is a healthy stimulant for academic enquiry. Diversity of opinion becomes a problem only when it prevents the application of systematic and consistent criteria for publication or research support and, by implication, for the advancement of a particular theory, area of study, or individual's career. In their study, Cole, Cole, and Simon found that the outcome of the peer review of any particular applicant was largely a matter of chance.

Peters and Ceci made a much more dramatic allegation. They concluded that the name and institutional affiliation of the author influenced the journal reviewers more than the content of the article. This serious charge is supported by other investigations. J.A. Rowney and T.J. Zenisek, at the University of Calgary, studied the reasons for the acceptance or rejection of manuscripts and found that the author's reputation was the most important determinant of reviewer support after the original and significant content of the manuscript itself.[37] An analysis of U.S. National Science Foundation program funding practices found that university-based research proposals were more highly rated than those based elsewhere.[38]

It is distressing enough that journals may allow reviewers to apply favouritism regarding who gets published. Even more disturbing is the possibility that when a manuscript contains ideas, hypotheses, or findings that coincide with the reviewer's beliefs, its chances of acceptance become far greater. Michael Mahoney, a psychologist at University of California at Santa Barbara, studied just how much a manuscript's findings and conclusions influence its chances of acceptance for publication.[39] He sent out seventy-five manuscripts to reviewers from a well-known psychological journal. All the manuscripts were anonymously authored and had identical introductions, methods, and bibliographies. However, in some cases, the results supported the presumed views of the referees; in others the results were contrary to them. Consistently, the articles supporting the referees' views were recommended for publication; the others were not.

This indicates that there is a more serious problem than the scientific snobbery alleged by Peters and Ceci. Mahoney's study leads us to the conclusion that reviewers consistently favour papers reflecting their own theoretical biases and scientific views.

Editorial licence to select a limited number of reviewers, who are free to operate without outside scrutiny, further compounds the individual biases of reviewers. Journal editors are usually scientists themselves who, like reviewers, often favour the familiar, the unadventurous, and the well-known, in both people and ideas. The established scientists generally selected as peer reviewers are much less likely to question the benefactors, theories, and subfields on which their careers are based than are young scientists who have not yet developed such strong allegiances. In this way, journals can become doctrinaire, deliberately eschewing challenges to accepted theories.

Ultimately, it is the editors of a journal who bear the responsibility to their readers for employing a diverse, intellectually open set of reviewers, and for periodically assessing the reviewers' performance. Such an approach is, however, exceptional, as Carl J. Sindermann observes in *Winning the Games Scientists Play*: "editorial boards of some journals can be closed circles, in which the editor selects those of like mind to serve as reviewers and in which members nominate compatible new members as replacements when terms expire."[40]

Anonymity further fosters selective access to the scientific press. A review system that relies on nameless reviewers can, all too often, result in arbitrary or blatantly biased decisions for which no one can be called to account. It becomes much easier for the reviewers to favour the applicants they know, or have heard of, or who share their scientific viewpoints.

Editorial decisions carry tremendous weight because there is no way for an author to appeal judgement on a manuscript, as Stevan Harnad, editor of *The Behavioral and Brain Sciences*, explains:

> There is no such thing as a "higher tribunal" of a different kind. Anonymous peers control the funding and publication of the work of their peers. But anonymity, as well as the closed conduct of the reviewing procedure itself (even though this clearly constitutes the optimal system), cannot help but infuse a conservative element into an enterprise in which originality and innovation are avowedly at a premium. This conservatism has great utility inasmuch as it enforces the constraints of consistency, testability and self-correctiveness in that large proportion of cases in which peer review has its intended effect. But human frailty (and indeed peer frailty) being what it is, there will be cases in which the conservative forces of peer review may underrate an effort that is of potential value, or may inadvertently create conditions in which some potentially worthwhile initiatives are stillborn. Second, in the very fact that the conduct

> of peer review is anonymous and closed, many
> aspects of the "creative disagreement" process are
> lost, and an unrepresentative impression of
> univocality takes the place of the much more
> diverse spectrum of alternatives, critiques and
> counter-examples that actually exist.[41]

The studies reported above show that the peer-review system does not necessarily select the best scientific projects for publication; nor is it always capable of recognizing and rejecting the worst. It may systematically support orthodoxy in experimental approach, in the method of data analysis, and in its interpretation. This tends to concentrate research effort and reporting in a few select subfields. Moreover, the validity of established research trails may rarely be questioned, because scientific replication is not greatly valued by journals eager to publish newer, more exciting results. Replication, the practice of repeating scientific work to make sure that the original findings hold true, is much touted as the guarantor of scientific truth.[42] But those who try to duplicate someone else's study are ill rewarded.[43] In a study of manuscript characteristics influencing reviewers' decisions it was found that "direct replications were of little interest".[44] An exact copy of a previous study contributes only the information that the original work was or was not valid, and it is, of course, much more challenging to build on existing knowledge, pursuing questions begged by past work. Understandably, alteration of one or two research parameters in a study might well render a report on it more acceptable than a direct replication. While such development of a research trail may reveal flaws in previous investigations, this won't always be the case. Derivative lines of research may thus be encouraged, and, under these circumstances, incorrect ones might persist for some time.

The review of applications for publication by a select group of peers inevitably politicizes science: groups of scientists who have a specific approach to science and who are represented on the peer-review committees are "in"; others are "out". The distribution of scarce funding and publication opportunities by the "in" groups helps perpetuate their views. As David

Horrobin, an independent biomedical researcher, has noted: "It is ... possible to destroy an application by deliberately sending it to eminent referees who are known to be opposed to the work."[45]

Scientific journals are essential to modern science. Dissemination of new results, hypotheses, and theories is crucial; without the specialized scientific periodicals, investigators would be infinitely more isolated and might easily waste time pursuing research trails that others had found to be fruitless. New theories of general relevance are discussed on the pages of the less specialized scientific publications; the current events that influence science policy can be debated there too, and these journals can act as fora for the development of different positions on scientific matters, or on public-policy issues. But all these purposes could be better served by a more diverse group of publications and an improved process for vetting manuscripts. New avenues for scientific communication should bypass traditional peer-review procedures to better circulate information on recent studies. Computer-linked data bases offer one such possibility; automatic entry of abstracts in an international data base already ensures that, in some fields, all studies, no matter how idiosyncratic, are disseminated in an abbreviated form.[46] Publications such as *Medical Hypotheses*, edited by David Horrobin, and *International Journal of Forecasting*, edited by J. Scott Armstrong and R. Fildes, also encourage consideration of novel or controversial ideas. If, in addition, public and private funding agencies allocated grants for reporting as well as for carrying out research projects, independent publication of controversial theories or results would be easier. Editors could be subsidized, by part-time salaries in the form of grants, to carry out the time-consuming review practices advocated below. Such funding should, however, be contingent on the definition and acceptance of clear editorial responsibilities and the development of guidelines for more acceptable review practices.

As a first step in this direction, all scientific periodicals should develop and enforce strict co-authorship policies limiting the list of authors to those making significant contributions to the work presented in a paper. Arnold Relman, editor of *The New England*

Journal of Medicine, explains what this would mean: "Coauthorship should denote at least that there has been meaningful participation in the planning, design, and interpretation of the experiments and in the writing of the paper. Multiauthored papers should never be submitted for publication without the concurrence of all authors. Before accepting a multiauthored paper for publication, editors would be well advised to assure themselves that these conditions have been met, perhaps by having all coauthors sign an appropriate form."[47] Daryl Chubin, of the Georgia Institute of Technology, suggests an additional safeguard: "[the inclusion of] an Acknowledgement section in each published paper indicating which author was primarily responsible for each component of the research would pinpoint accountability without undermining the collaborative spirit."[48] Senior scientists might then be prevented from automatically adding their names to the papers published by researchers in their labs, and students and more junior scientists would get proper credit for their work. (Laboratory hierarchies were analysed in greater detail in Chapter Five.) Another measure that would act to limit a senior scientist's prestige when it is excessively based on the research efforts of others, would be to present publication lists as fractions: the number of papers divided by the number of authors.[49] If this indicator could be adapted to reflect the degree of involvement of each author, it would be a much more accurate estimate of the individual's contribution than would the total number of papers bearing his or her name.

Another initiative for promoting the just attribution of credit is the suggestion (followed by *The Behavioral and Brain Sciences*) that the usefulness of reviewers' comments be rated by authors, on a simple form, for the benefit of editors.[50] Anticipation of this review might temper the worst excesses of negligent or cavalier referees. Reviewers who are thorough, constructive, and fair would be rewarded (perhaps with some published recognition of their services), while those who use their position to vent their spleen or prejudices would be exposed. If, in addition, authors were given a more detailed explanation of the decision made on their submission,[51] and the right to respond to it, these measures might moderate, at least partially, the current situation in which "authors are tried and condemned by anonymous judges without being invited to defend themselves".[52]

Michael Mahoney suggests another way to promote reviewer efficiency and give authors a little more power in the publication process. He proposes the simultaneous submission of manuscripts to several journals. This is currently banned as causing redundant reviewer effort, and could result in complex legal undertakings for both authors and publishers. Mahoney contends that simultaneous submissions would, none the less, speed up the peer-review process, while involving a similar number of reviewers as those employed in the sequential review system, in which a manuscript is consecutively submitted to less widely read journals, until one finally accepts it.

Another measure that would improve the quality of publication and help to reveal fraudulent work is the insistence that data books, including laboratory protocols and results, be retained and made available for inspection for at least five years after completion of the research.[53] A detailed Methods section (including, for example, all relevant chemical formulae and technical procedures) should be published or, where length does not permit this, filed with the editors, as is suggested by *International Journal of Forecasting*.[54] This would ease accurate replication of the work presented. (It would also preclude the all too frequent practice of withholding complete methods sections in order to discourage intrusion into a promising new research area.)[55] Such conditions for publication could be laid out in guidelines printed in each journal.

The likelihood of plagiarism and reviewer bias could be substantially reduced by choosing reviewers who do not have a personal investment in the research addressed by the manuscript. For instance, reviewers suggested by authors and investigators outside the author's specialized subfield or geographical area would help to achieve balance.[56] All referees could be given detailed procedures for manuscript review, and cautioned to look out for "malpractice in data manipulation or interpretation".[57] Stevan Harnad advocates the use of computer search methods and large current bibliographic data bases to assist in fairly selecting expert reviewers.[58] In addition, if time and money permit, several reviewers (perhaps more than the two or three currently used by most journals) should be consulted for each manuscript submitted, to dilute the impact of one unfair review.[59]

Peters and Ceci suggest that making the authors of articles anonymous, like the reviewers, would reduce favouritism and discrimination, by forcing reviewers to judge an article on its intrinsic merits. But such a blind review process is no panacea. Sophisticated reviewers familiar with a subfield could easily identify anonymous authors. In addition, this double-blind technique doesn't address the most serious issue, that of reviewer bias regarding the findings or theoretical bias of a paper, even if the author is unknown.

There are two contrasting ways to resolve this problem. Mahoney suggests that papers could be submitted with their introduction and methods sections intact, but with their discussion and conclusions omitted: reviewers would have to judge manuscripts that are "blind" regarding their conclusions.[60] A modified form of this procedure is occasionally practised by *International Journal of Forecasting* as part of its larger effort to publish important, surprising, or controversial work.[61] Mahoney's "blind" review method must, however, be time-consuming for reviewers and could render editorial judgements complex and difficult. A better alternative might be for authors and journal editors to acknowledge the influence of their own values, ideas, and affiliations in their work, and to reveal them without embarrassment. Ian Mitroff, a scholar concerned with peer review, proposes a daring format for scientific publications:

> Each page of the paper would have a line down the middle. On the right half of each page (perhaps corresponding to what is currently being attributed to the left half of the brain) would appear the standard, traditional account of the most tightly controlled inquiry I am capable of doing. On the left half (right brain) would appear a blow-by-blow, stream of consciousness account

of what I thought, felt, and so on, as I went through the pain, sweat, and joys of doing the study. There would be no attempt to tie these two sides together. They would merely sit there together, existing with and without the other. No comment would be given.[62]

But solutions to peer-review problems are not easy. Mitroff's suggestion would considerably lengthen articles, creating even greater competition for the reduced publication space available. Furthermore, his proposal would likely make the theoretical preferences of journal editors much more obvious, and could, as a result, lead to the polarization of journals supporting certain theories or approaches. Moreover, the open debate of important underlying assumptions in widely read publications with less partisan outlooks could easily be discouraged.

The argument that scientists' underlying values are an important influence on the way their papers are formulated and perceived leads to another, more practical suggestion – "reviewer disclosure". In this reversal of blind peer review, reviewers would routinely sign their names to their critiques of manuscripts. Such openness might curb the unkind or unfair statements that proliferate under the present system. Not surprisingly, this is not a popular proposal with reviewers. Sixty-eight per cent of the Canadian Psychological Association reviewers surveyed by Rowney and Zenisek favoured continuing the present system.[63] Critics of reviewer disclosure protest that it might well inhibit their work. Young reviewers might be reluctant to criticize the work of a prestigious and influential scientist, and reviewers might simply refuse to critique manuscripts.[64] Nevertheless, the abstention or self-censorship of timid reviewers seems far preferable to the potential for favouritism, discrimination, and repression of intellectual challenge inherent in the current peer-review process.

An extension of reviewer disclosure is the practice of open peer commentary followed by the journals *Current Anthropology* and *The Behavioral and Brain Sciences*. These journals publish articles (the acceptance of which is determined by peer review) along with a series of commentaries on the article by colleagues. In this way, each issue of the journal becomes a forum for debate on the subjects raised in the main paper. Authors' biases are implicitly recognized and counteracted by those of the commentaries, and the field is enriched by numerous contributions. Clearly this couldn't be a regular procedure for all journals, or each paper would generate an entire issue. This would lead either to a proliferation of journals to meet the increased demand to publish or to even greater competition to be published in the limited number of existing journals. Furthermore, only some pieces of work are provocative enough to excite a broadly varied response from commentators. Efforts to harness peer dissent into a creative force are a welcome response to the problem, but an open peer-commentary system can be practical only in particular circumstances. Inevitably, for the majority of publications, a more restrictive publishing practice will have to be maintained.

One doesn't have to be too cynical, however, to see that reviewer disclosure combined with reviewer rating would go some way towards checking systematic favouritism and prejudice, although bias against ideas and theories could persist. A more profound change is needed to recognize the role of personal bias in scientific research and to assess its products accordingly. A growing number of scientists acknowledge that their opinions, experiences, and values inevitably affect their judgements, and they are designing their courses on research method to reflect this admission. But the image of their profession as dispassionate, reliable, and expert probably appeals most to scientific researchers themselves – collectively they may be the last to question their own traditions and practices.

There is, therefore, urgent need for a fundamental reworking of non-scientists' expectations of scientists and of their view of science, as a dispassionate search for an "objective" reality. Scientific funding is predicated on this same notion, and would require similar restructuring to reflect the inevitable emotional and intellectual baggage that scientists bring to their work. Just how research grants are currently acquired and maintained is examined in more detail in the following chapter.

Chapter 7

FINAGLING AND FUNDING

"I have earned a very large sum of money," said Mr. Gable.... "And ... I want to do something that will really contribute to the happiness of mankind...."

"Would you intend to do anything for the advancement of science?" I asked.

"No," Mark Gable said. "I believe scientific progress is too fast as it is."

"... [T]hen why not do something about the retardation of scientific progress?"

"That I would very much like to do...."

"Well," I said, "... You could set up a foundation, with an annual endowment.... Research workers in need of funds could apply for grants.... Have ten committees ... appointed to pass on these applications. Take the most active scientists ... and make them members of these committees. And the very best ... should be appointed as chairmen.... Also have ... prizes ... for the best scientific papers of the year...."

"First of all, the best scientists would be removed from their laboratories and kept busy on committees.... Secondly, the scientific workers in need of funds would concentrate on problems which were ... pretty certain to lead to publishable results..... [P]retty soon science would dry out.... There would be fashions. Those who followed the fashion would get grants. Those who wouldn't ... would learn to follow the fashion, too."

-- Leo Szilard, "The Mark Gable Foundation", p. 99.

This cautionary tale is recounted by Leo Szilard, scientist and writer, in his book *The Voice of the Dolphins and Other Stories.*[1] He, like most modern investigators, was uncomfortably aware that the system of research funding is the lifeblood of science – the strongest single determinant of its current organization. This system fosters and even requires a hierarchy within the community of scientists. More insidious is the influence that grant-giving committees can wield over the fields, specialties, and hypotheses investigated. That pervasive power and its exploitation are the subject of this chapter.

Long gone are the times when individual researchers such as Darwin, Newton, and Faraday, working alone, produced dramatic results through simple observation or experiments. Nowadays, writing, reviewing publications, serving on academic and professional committees, speaking engagements, lecturing and supervisory commitments eat up most of the time of senior scientists. A hierarchy of middle and junior scientists, post-doctoral researchers, graduate students, and technicians maintain research production in large laboratories. In short, science has become bureaucratized.

Securing and maintaining research grants to buy expensive equipment, pay research staff, and maintain the university premises is the laboratory's paramount concern. Senior scientists prepare grant applications, defend the results of previously funded work, and serve on committees that dispense money. This system can be preposterously inefficient. As Rustum Roy, Director of the Materials Research Laboratory at Pennsylvania State University, notes, in one recent case it cost at least $16,000 in scientists' time to evaluate and process applications for each grant of $30,000.[2]

This situation makes investigators extremely dependent on granting committees and consequently gives those on the granting committees the power to decide who, and what work, is worthy of support.

Most scientific-research grants are handed out for government agencies by committees of scientists appointed to review applications in a particular research field.[3] They are asked to evaluate grant applications on the basis of several criteria. The following, used by the National Institutes of Health in the United States, is one example:

> (i) the scientific, technical, or medical significance and originality of the research; (ii) the appropriateness and adequacy of the experimental design and methods; (iii) the qualifications and experience of the investigator(s); (iv) the reasonable availability of resources; (v) the reasonableness of the proposed budget and duration of support in relation to the proposed research; and (vi) where an application involves activities that could have an adverse effect upon humans, animals, or the environment, the adequacy of the proposed means for protecting against such effects. NIH considers these criteria to be sound and sufficient to enable the peer review system to identify the most meritorious research grant applications for support.[4]

Several studies have investigated how the granting system influences who and what get research support. One of the most comprehensive explorations was commissioned in the late 1970s by the United States National Science Foundation (NSF) on behalf of the National Academy of Sciences. Stephen Cole and Jonathan Cole, in conjunction with Gary Simon and L. Rubin, examined the way that more than 250 grant proposals were rated by the appointed NSF referees. They also compared the decisions made by NSF on 150 other proposals with those rendered by different, independently selected panels of experts. Their conclusions were startling. They found that, although the NSF review ratings were the primary determinant of an applicant's success, the reviewers disagreed so dramatically among themselves that it was largely the random chance of having favourable reviewers that determined whether an applicant received funding. The report did not, however, conclude that there was *systematic* bias in the selection of NSF reviewers or in the ratings of those reviewers, based on the age, affiliations, or achievements of the applicants.[5] This, despite their assertion that "scientists with an established track record, many scientific publications, a high frequency of citations, a record of having received grants from the NSF and ties to prestigious academic departments have a higher probability of receiving NSF grants than other applicants do."[6]

Their finding that "the granting process is actually quite open, and there is nothing approximating a scientific caste system" has been challenged by I.I. Mitroff and D.E. Chubin,[7] two social scientists who have been studying how individuals involved in scientific research perceive and conduct their work. They questioned the Coles' conclusions and wondered just how willing they were to embrace more critical interpretations. Mitroff and Chubin criticize Cole, Cole, and Simon's intepretation of the data.[8] They charge that the Coles equate quality with frequent citation of work and fail to analyse how the affiliation of the applicants coloured the assessment of their proposals by the reviewers. They are also critical of the sample, which, they assert, was not balanced. Finally, they judge Cole et al.'s efforts in this way: "Though they began with good intentions, Cole and his associates may have done more to defend the status quo than to inform the debate on peer review...."[9]

Other empirical investigations more directly address the fairness and wisdom of the current granting system. I looked at the membership of the Medical Research Council of Canada (MRC) grant-award committees that operated during the period 1971-81. I found that these committees are rather small – between six and eleven scientists on each of nineteen committees, with an average turnover of two or three members per committee per year. (Committees are aided in their assessment of applications by between one and three external reviews of each. The final judgement on each proposal, however, rests with the committee.) These two hundred or so committee members represent only a fraction of the sixteen hundred or so researchers funded each year, and an even smaller proportion of the four thousand applicants. (During this period, typically 30 to 35 per cent of new grant applications and 87 to 90 per cent of grant-renewal applications were successful.)[10] Moreover, several individuals served on more than one committee during the ten-year period, or served on the same committee for more than one term. I also examined the composition of editorial boards of several Canadian medical research journals listed in *Index Medicus,* and found that nearly every editorial board included scientists who were or had recently been on an MRC grants committee.

Having established that a relatively small number of research scientists controlled the funding and publication decisions in their fields, I then looked specifically at which scientists and what proposals were receiving support. Like other granting agencies, the MRC advises grant committee members to absent themselves from discussion of their own colleagues' applications. Nor do they see the reviews of their own grant applications before anyone else. Nonetheless, their usual participation on the committee, and their familiarity and credibility in the eyes of their fellow committee members may give them some advantage. Indeed, I found that the committee members do very well by the MRC. The average MRC grant in 1979-80 was $34,041. The average amount given to a grant-committee member during that same year was $52,376! Furthermore, all but one of the recipients of the ten largest MRC grants were on grants committees during the years 1971-81. In 1979-80, 73 per cent of all grants of more than $100,000 and 65 per cent of all grants of more than $75,000 were awarded to scientists who had been on an MRC grant committee at some time during 1971-81.

It seems that the projects given the most generous support are those proposed by the very people who sit on the committees. It may well be that the best scientists propose the most worthwhile projects and therefore get the most support, and similarly that the most experienced and far-sighted individuals are recruited for the grants committees and journal editorial boards. But it can equally be argued that this process of peer review is inherently conservative. Investigators who adhere to the prevailing dogma and work in established fields with respected scientists are most likely to receive funding and get their work published. As their reputation develops, they likely continue to receive a disproportionate share of the research pie, making it increasingly difficult for less well-known researchers with unorthodox ideas to get support.[11]

A report detailing the systemic discrimination against female researchers, prepared by David Horrobin, provides further support for this theory:

> A number of women complained to the Modern Language Association in the United States that there were surprisingly few articles by women in the association's journal, compared to what would be expected from the number of women members. It was suggested that the review processes were biased. The association vigorously denied this but under pressure instituted a blind reviewing procedure under which the names of the authors and their institutional affiliations were omitted from the material sent to the reviewer. The result was unequivocal: There was a dramatic rise in the acceptance of papers by female authors.[12]

The power vested in grant-giving committees clearly enables them to favour friends, mentors, and potential benefactors, and to discriminate against competitors or advocates of theories that the committee members do not support. While many scientists feel that most grant-givers are usually fair and dispassionate, charges of favouritism and discrimination such as those levelled by the above studies are gaining acceptance as research funds become scarcer.[13] The authors of a review of the research evaluation in British science find that the peer-review system used there has three main weaknesses: it has allowed certain disciplines and institutions to become entrenched in the granting process, thus ensuring secure funding for their own; research requiring expensive equipment has become increasingly concentrated in a few well-endowed centres; and budget cuts have forced reviewers to choose which work will *not* be funded, a decision rarely made on grounds of scientific merit alone.[14]

Deborah Hensler, in a study of the U.S. National Science Foundation in 1975-76, confirmed that most reviewers and applicants believed that older, well-established applicants at well-known institutions, whose proposals were consistent with the mainstream of thought, had the best chance of being funded.[15] A more recent study of the NSF also found that "reviewers favoured basic scientific research conducted in an academic setting", and that they had a strong preference for proposals originating in their own field.[16] Ernest Borek, a professor of microbiology, accuses grant-givers of "bias and dedication ... to the defense of their research turf", in an editorial in the journal *Trends in Biochemical Sciences*.[17] Rustum Roy addressed the 1984 meetings of the American Association for the Advancement of Science, likening peer review to "a jury of axe murderers from the same gang".[18] He is paraphrased in *New Scientist* as follows:

> There is no attempt to weed out reviewers with built-in prejudices, or who receive funds from the same source as an organisation for which they are reviewing proposals for research grants.
>
> Such people have a vested interest in turning down proposals because in that way they can preserve funds from which they draw their own support. And who will ensure that a researcher does not vet research proposals from a bitter enemy? Roy believes that everyone who reviews research proposals should sign a statement that they have no conflict of interest.[19]

The grant-giver's power to reward friends and punish enemies is extended to those receiving hefty research grants. These recipients are able to employ favoured students and colleagues. If scientists' grants are mobile, they can be used as negotiating tools when changing jobs or departments. The portion of a research grant that goes to a researcher's home institution as overhead adds considerable appeal to a prospective employee. On January 29, 1981, *New Scientist* revealed how the political adviser to the British Secretary of State for Transport, Ian Heggie, attempted to use his prospective Department of Transport grants to persuade Oxford University to offer him a prestigious position.[20] Preferential treatment can percolate from grant-givers to favoured applicants, to the applicants' students, creating a scientific élite of established or soon-to-be-established researchers, who have security and the power to confer it on others. J.M. Ziman, the eminent Professor of Theoretical Physics at Imperial College of Science and Technology, University of London, England, describes the process of gaining entry to this élite group, which he calls an Invisible College:

> How, for example, does one become accepted as a charter member of an Invisible College? It is not sufficient merely to be doing research in that field, or to publish a paper on it, unless this is of such extraordinary merit as to catch the attention of the whole community. The usual entry is achieved by patronage. The students of a particular professor are recommended to his colleagues not merely for jobs, but as potentially able contributors to the field. To have taken one's doctorate in some famous school of research provides one with a ticket of admission, as a visitor or temporary research worker, to another distinguished group, where one's name will then become known. A joint paper with a leading scholar may be mainly the work of the student – but the name upon it may provide the necessary *cachet* for his further advancement.[21]

Even more disturbing than systemic favouritism and

discrimination is the possibility that particular fields or scientific hypotheses may be consistently favoured over others. A striking and alarming example of this is the dramatically disproportionate investment in U.S. military research and development (R&D) compared to civilian R&D, particularly research on social needs and problems ("between 1980 and 1983, [U.S.] federal spending on defence-related R&D increased by 22.3 percent in real terms; in contrast, federally supported civilian R&D decreased by 30 percent over the same period").[22]

In behavioural research there is a similar trend to support certain pursuits, regardless of their public utility or their destructiveness: male-only subjects are used in a great many studies, yet the results are frequently generalized to apply to women as well. Moreover, despite the disproportionate number of women receiving psychiatric care, only a very small fraction of the Canadian grants awarded for mental-health research went to projects directly related to women.[23] Because this tradition is so well established, it has a veneer of acceptability despite its obvious discrimination and potentially misleading consequences.

The distressing possibility that new approaches to old problems are unwelcome, even if they prove productive, is supported by a large body of diverse evidence. Bernard Barber, a sociologist of science at Columbia University, is one of the many investigators who have documented (in an article in *Science*) historical resistance by scientists to scientific discoveries.[24] From Copernicus in the early sixteenth century, to Mendel, Pasteur, and, of course, Darwin in the nineteenth century, new theories and observations have been greeted with alarm and hostility. The frustrations caused by this rejection led the eminent physicist Max Planck to the bitter conclusion that "[a] new scientific truth does not triumph by convincing its opponents and making them see the light, but rather because its opponents eventually die, and a new generation grows up that is familiar with it."[25] Leo Szilard, who was a prominent U.S. civil engineer and scientist, is reported to have been so distrustful of grant committees' willingness to support truly original work that he applied for funding only for work that he had already completed. (On one occasion, he was refused support because the work was considered impossible!)[26] The difficulty of securing funding for work of dubious outcome is widely recognized,[27] and Szilard is not the only one to have chosen this less than honest, if effective, solution.[28] Rejection of innovative or unusual research methods or repression of novel research specialties have been described by scientists in fields as diverse as ecology, geology, and biochemistry.[29]

The resistance to new ideas allows the undue persistence of old ones.[30] Research support is generally much easier to secure when the investigation proposed develops an established line of study. Indeed, scientists beginning their careers bear the added burden of having unfamiliar names and, if they are prudent, they will make their research choices with care. (This was discussed in greater detail in Chapter Six.) Young investigators who are experienced enough to be cynical, yet dedicated enough to pursue a scientific career, will know that their ability to secure research funds on their own is limited. They will be aware that, instead, they should cultivate well-known supervisors and departments, and work in fashionable research fields, which promise early returns on research investment. Failure to choose shrewdly might easily mean that a promising but naive young scientist is excluded from consideration for many positions early in his or her career.[31] In a report published in *Sociology of the Sciences*, D. Chubin and T. Connolly[32] point out that these unwritten rules inevitably favour conservative students, willing to further an established field, over those who attempt original work in an untested specialty. The current science-funding system dictates that the most secure investment for the ambitious academic is plodding investigation of a narrow project that incrementally develops an established research trail.

The institutionalized favouring of old ideas, fields, and research directions can be explained in part by the different standards of evidence applied. We require much less conclusive proof to buttress an accepted theory than to establish a radical new one.[33] (This was discussed in greater detail in Chapter Three.) An anonymous research-institute member and director of the National Academy, quoted in the book *Politics in Science* by Marlan Blissett, a professor at the Lyndon B. Johnson School of Public Affairs in Texas, put it this way:

> In my profession there is a canonizing of views. These are the views that are subscribed to by the most influential members of the profession. That is, those who are the editors of journals and the leading members of professional societies. These people are the defendants and judges of particular points of view. They are able to crush incompatible thoughts. But, of course, they do it under the guise that there is not enough evidence, or that the evidence is incomplete. You might say that there is a conspiracy of discrimination. Why? Because some people fear that their reputations are threatened or they fear for their economic position within universities. A new theory or innovative model creates certain insecurities.
>
> In general I would say that the barrier to the advancement of science lies between the whole of a discipline and the in-group that controls the flow of discoveries. What happens is that there are flag-wavers in science as there are in the larger political system. In science the flag is not Old Glory, but "complete evidence." And many scientists turn themselves into apostles of complete evidence.[34]

In short, science has become politicized: the ideal of rewarding the originality, promise, or rigour of a proposal can all too easily be supplanted by familiarity, support for one's own theories or those benefiting powerful interests, and recognition of the applicants and their home institutions.

Such intellectual conservatism is not unfamiliar. In business, politics, and social affairs, participants are expected to change their preconceptions only with great reluctance. Yet scientists like to think that science should be immune to traditional conformist pressures. Unlike many other professionals, scientists have developed rules for their experimental work to curtail the influence of investigators' desires on their findings. But these pressures may be psychologically inescapable.

In numerous publications, psychologists Robert Rosenthal, L. Ross, N.E. Nisbett, and many colleagues have demonstrated that people are highly biased in the way that they search for, recollect, and assimilate information. They show that these processes operate to bolster one's original beliefs and expectations. Experiments became self-fulfilling prophecies,[35] and even when the initial basis for a belief is invalidated the belief may be sustained.[36] In one of their experiments, C.A. Anderson, M.R. Lepper, and L. Ross showed subjects pairs of fictitious examples that related firefighters' job performance to their willingness to take risks. The subjects seemed able to rationalize a relationship between good firefighting and foolhardy or cautious behaviour, based on the attitudes of the individuals described. Subsequently their faith in the relationship continued unabated, even when they were informed that the original attitudes described, which had brought them to this view, were fabricated.[37]

These psychologists and others working in the field of "cognitive dissonance" theory have speculated about the mechanisms promoting undue persistence of existing beliefs. First, memories consistent with the initial belief are marshalled to support it; second, people try to explain new beliefs by attributing plausible causes to them. When the original evidence for the new belief is subsequently destroyed, the congruent impressions and possible explanations for the belief assembled to support it remain. The individual holds on to the belief even after it has been shown to be groundless.

From this fascinating psychological explanation for people's reluctance to let go of old dogmas and accept new ones we may infer that, throughout evolution or generations of consistent developmental training, intellectual conservatism has been productive most of the time. Even if inaccurate, a long-held theory may well provide useful predictions; occasional contradictory evidence should be viewed as exceptions to a generally sound rule. For, if every trivial item of new knowledge prompted a complete re-evaluation of long-accepted theories, we'd be hard-pressed to learn anything; our minds would be in chaos, wondering what, if anything, could be accepted as true.

Some scientific orthodoxy is helpful; it encourages a disciplined approach to research, and allows us to compare different studies. A widely accepted status quo simplifies the assimilation of new and congruent information, and gives scientists in different settings access to a common set of theories or an agreed structure for organizing knowledge. However, as the eminent philosopher of science Michael Polanyi has noted, the dangers posed by powerful interests vested in the current dogma should not be minimized:

> I am not arguing against the present balance between the powers of orthodoxy and the rights of dissent in science. I merely insist on acknowledgement of the fact that the scientific method is, and must be, disciplined by an orthodoxy which can permit only a limited degree of dissent, and that such dissent is fraught with grave risks to the dissenter. I demand a clear recognition of this situation for the sake of our intellectual honesty as scientists, and I charge that this situation is not recognized today but is, on the contrary, obscured by current declarations about science. Take this by Bertrand Russell: "The triumphs of science are due to the substitution of observation and inference for authority. Every attempt to revive authority in intellectual matters is a retrograde step...." Such statements obscure the fact that the authority of current scientific opinion is indispensable to the discipline of scientific institutions; that its functions are invaluable, even though its dangers are an unceasing menace to scientific progress.[38]

The weaknesses of the system for funding scientific research have been increasingly recognized in recent years, although this recognition has yet to be translated into widespread reforms. There is, however, an active debate about how best to improve

141

current grant procedures. Investigators at the Science Policy Research Unit in Sussex, England, support the use of more objective, quantifiable measures of research performance.[39] So does Rustum Roy, who has developed a new numerical measure, calculated on the number of graduate degrees awarded, the number of papers published, and the extent of industrial research support and targeted grants, funding applied to solving a particular problem, previously awarded to a research team or department. Increases in any of these factors would produce a corresponding increase in the rating figure.[40]

Roy's proposal would, however, address few of the science-funding system's problems. The concentration of research funds in the hands of established researchers, teams, and institutions, now accomplished by favourable subjective reviews, would be further promoted by the undue weight given to the number of publications and the faith placed in the team by other granting bodies. Researchers would be encouraged to work on industrial or goal-oriented targeted projects, as these would increase the level of government support. Research challenging corporate interests would get short shrift, as would investigations that have no apparent industrial relevance. (Industrial funding of academic research was discussed in greater detail in Chapter Four.) Because a larger volume of publications and qualified post-graduates generated by an establishment would inflate its grant, trivial articles would be encouraged, as would reduced standards for graduate degrees. Quantity would triumph over quality, and the older and more orthodox research élite would in no way be threatened.

Nor would stricter guidelines for politically determined, "targeted", or "mission-oriented" research solve these problems. Applied research on topics urgently demanding attention should not be discouraged, but practical advances for human well-being are often based on earlier "pure" research, which tends to be motivated by curiosity, not social needs.[41]

I would suggest, instead, a more egalitarian granting system. Peer review may be the best of a bad lot of options, but its current practice can certainly be improved. Granting agencies should experiment with a variety of reforms; their merits will only be revealed by trying them out. Some of my own proposals, as well as those of others interested in peer review, follow.

For the majority of fields, which do not require large, multi-faceted research efforts, more and smaller grants could be divided among larger numbers of investigators. Proven research ability should not, of course, be penalized; but neither should its rewards engender an exclusive research establishment. Unlike the current trend to invest in "centres of excellence" with an established record of research success, dispensing a larger number of smaller grants would discourage the development of enormous research empires while encouraging a greater variety of avenues of investigation. Major equipment grants could be provided on a departmental basis, and decisions on their use could be made co-operatively rather than by a single investigator.

Most important, the reviewers should be selected from a more catholic body of researchers, including young investigators at lower-status institutions and independent freelance researchers. At least in fields where the research topics bear on matters of public interest, these academics should be joined on the grant committees by reviewers drawn from the interested public and the communities benefiting from the research.[42] While this would, no doubt, be difficult for highly complex areas of pure research such as, for example, particle physics, in many other fields it would greatly improve the process. The U.S. National Institutes of Health uses a two-stage grant-review procedure whereby a scientific committee assesses the academic merit of each proposal, and a more diverse committee including non-scientists considers other factors, such as the total amount of funding available and the priority given to the particular research field.[43] More generally, nurses, paramedics, public-health advocates, patients, and hospital administrators could be involved in granting decisions concerning medical research, environmental, consumer, and other public advocates in other science and engineering investigations. If these lay members have sufficient experience in the subject area, they should have the self-confidence and critical ability to temper the prevailing academic conservatism.[44]

Applicants could also be given an opportunity to reply to reviewers' critiques of their proposals before final granting decisions are made.[45] Conflict-of-interest guidelines should be developed to advise reviewers when to abstain from judging proposals. This would be especially important if lay members serve on granting committees, in order to prevent businessmen's commercial self-interest from directing publicly funded research effort. Granting agencies might consider establishing an "explorations" or "innovations" grant scheme, for which reviewers would be specifically directed to place a high priority on interdisciplinary investigations that transcend the disciplinary constraints of most granting bodies, original hypotheses, novel research trails, and new or different analyses of existing data.[46] Together, these measures could significantly cut down on reviewers' ability to promote friends and favoured theories while frustrating their competitors.

It is clear, however, that restructuring the granting system alone will not counteract scientific prejudice or scientific conservatism. Changes in funding must be accompanied by parallel reforms in publication practices, by the deliberate dismantling of large research empires, and by the creation of more egalitarian laboratory structures. Together, these complementary changes could have a profound effect. This conscious disengagement of science from the vested interests it serves should result in a more responsive, flexible, and productive research enterprise.

CONCLUSIONS

A number of important ideas emerge from the preceding chapters. These ideas are not new: their theoretical implications have been debated by historians and philosophers of science for decades.[1] But they have a particular current relevance for the practical administration of scientific research.

We must begin by acknowledging the sheer number of choices that face research scientists, a number that grows with each new methodological development and analytical innovation. Not only must investigators select which field to explore and which problem to tackle, but they must also decide which assumptions, methodology, and analysis to apply to it, and how then to interpret and report their findings.

Inevitably, these crucial choices flow not only from the scientific traditions and habits that ensure experimental comparability and logical research sequences, but also from the personal preferences of the scientific investigator. Career pressures, academic allegiances, corporate patronage, and political sympathies can all contribute to a researcher's inclination to favour a certain conclusion. Regardless of the motive, however, a host of practices can steer the study towards desired findings. I have provided a glimpse of the variety of activities that may result in the dissemination of slanted conclusions or doubtful information. It may be helpful, however, to list points in the range, from the most reprehensible fraud to the mere avoidance of unwelcome interpretations. (The last four points describe practices that are also dishonest but do not cause circulation of dubious results; they are offences against people rather than against science itself.)

- Invention of entire experiments, complete with fictitious results.

- Invention of data.

- Alteration of data.

- Suppression of inconvenient data, either by omitting specific data points from a graph or report, or by failing to report an entire experiment.

- Suppression of unwelcome projects, hypotheses, or findings by unwarranted rejection of manuscripts or grant applications.

- Designing an experiment so that its results are inevitable and do not test any hypothesis.

- Adoption of invalid or dubious assumptions that bias experimental results or interpretation; failure to retract publications of work that relied on assumptions now known to be invalid or dubious.

- Analysing experimental results so that they appear to point in a predetermined direction.

- Interpreting experimental results in a way that supports a particular theory, without exploring alternate interpretations.

- Appropriation of the research data produced by others for personal gain.

- Presentation of others' data, analysis, or ideas without credit.

- Systematic discrimination against particular individuals or institutions, and favouritism of others.

- Republication of findings for personal gain without reference to their previous publication.

These activities range from the rare, conscious meddling with results to biases in research reporting that are widespread and likely inadvertent in all but the most introspective investigators. There is no clear delineation of what is deliberate and what is not; the most cold-blooded, opportunistic scientist might expressly develop an experimental protocol bound to produce results favouring a desired conclusion; sloppy scientists might rationalize a failure to report aberrant data points publicly in the literature as an artifact of methodological inconsistencies.

Researchers' awareness of their own motives and the reasons for their scientific choices will always depend on the personality and ethics of the individual. In *Betrayers of the Truth* William Broad and Nicholas Wade[2] argued that there was no difference between stretching the truth on the basis of faith in a theory when you turn out to be right (as Mendel and Newton did) and when you're wrong. They were widely criticized for likening Mendel and Newton, great scientists who may have exaggerated the strength of their data, to Burt and Morton, who concocted or misread evidence to support their now discredited theories.[3] Yet the important ethical distinction is not the correctness of the preferred theory but whether the scientists consciously intended to skew the findings. Morton probably didn't realize that his choice of method was responsible for his finding (that white cranial capacity exceeded Indian, which exceeded black).[4] He was innocent, but wrong. On the other hand, Millikan's laboratory notes show that he seemed deliberately to discard data that didn't support his theory. His work is important in the development of subatomic-particle physics, but in his published research reports he seems to have been intentionally deceptive.[5] There is no comfortable coincidence between the scientists' innocence and history's assessment of their contribution to knowledge.

Personal influences on scientific practice are neither isolated nor recent phenomena. They are a natural result of the human creation of scientific-research activity; work carried out can never be completely separated from the people who do it, or from their hopes, desires, and viewpoints. What has changed, over the last couple of decades, is the diversity and intensity of the pressures on scientists. Scientific evidence has become fundamental to public-policy decisions: political, corporate, and academic vested interests require data to support their positions. As a result it has become difficult if not impossible to distinguish a researcher's preferences from the ideology surrounding the research endeavour. The administration of science – the system of training, advancement, publication, and funding – is itself slanted towards certain fields, hypotheses, and individuals.

The authority structure within science makes it likely that pressure arising from particular vested interests will prevail over others. The peer-review process, which currently controls most scientific funding and publication decisions, produces what the philosopher of science Stephen Toulmin calls a "gerontocracy"[6] – the Old Boy network of successful scientists with power to determine research priorities, promote favoured theories or hypotheses, and make or break academic careers.

As government research funds shrink and competition for research money, jobs, and publishing space intensifies, these scientific power brokers will increasingly favour the familiar, established research trails and investigators. Erwin Chargaff, the noted Columbia University biochemist and science commentator, laments this trend in his article "In Praise of Smallness – How Can We Return to Small Science":

> As in every decaying commonwealth, so also in science, the financial powers and the means of production, that is, the laboratories, etc., have be-

come concentrated in ever fewer hands. We are
sailing straight into a managerial dictatorship in
which the individual scientists can no longer have
a voice. In other words, science has become
thoroughly politicized, a playball of power net-
works in which such expressions as "the search for
truth" or "the benefit of humanity" must sound un-
convincing, and even ridiculous, since everybody
knows what and who are behind them.[7]

If this system is not reformed, young researchers with novel
ideas or politically inconvenient findings will be increasingly
excluded from scientific activity. Ultimately, many traditional
dogmas will, no doubt, be overturned by more powerful
explanatory theories or productive research hypotheses; but, in
the meantime, the public will have been denied potentially
valuable new discoveries, inventions, or evidence challenging
current political practices. Vast sums of money may be
squandered on fields that are high political priorities but which,
like military research, are socially destructive and have few if any
incidental benefits.

As Michael Polanyi asserts,[8] science must always represent a
balance between theories that have worked in the past and
challenges to those theories that may prove more productive in the
future. The scales have tipped too far towards the traditional;
reform is urgently needed to diversify scientific endeavour and to
dethrone the power brokers who profit from the status quo and
would prevent constructive changes in scientific administration.

C.J. List, an American philosopher of science, argues that the
internal rewards of science, such as the aesthetic pleasure and joy
that come from exercising the intellect, must be emphasized, as
researchers with these motives will not be vulnerable to the
temptations that contribute to fraud.[9] I would add that the
external rewards of scientific success, the fame, if not the fortune
achieved by the most eminent researchers, must be reduced
accordingly. While such reforms might dismay the most
ambitious and opportunistic researchers, other perks, such as light
undergraduate teaching loads, the relative independence and

security provided by the tenure system, and the gratifications and publications accruing from graduate teaching, should all serve to maintain the appeal of academic, as opposed to industrial, positions. Improved university and research funding is essential in attracting and keeping excellent researchers who would otherwise gravitate to better-equipped and more spacious facilities elsewhere. Thus, initial steps should be taken, without delay, to create a better funded but more egalitarian, productive, and diverse scientific enterprise.

Two simultaneous approaches are needed to reform the way science is currently practised. First, we should all accept that bias unavoidably alters scientific research, and we should develop measures to challenge the biases of investigators. Second, we should take steps to disengage, as much as possible, the scientists doing research from the vested interests that have a stake in its outcome. These two approaches may appear to be contradictory, but they are not. Both recognize the inevitability of vested interests affecting research topics, methods, and results. The initial response is to foster awareness of this fact on the part of the researchers and to institutionalize challenge and examination of bias as part of the research process. A complementary program would operate at the level of departmental, institutional, and governmental policy, to dilute the impact of the vested interests on scientific investigation. The rewards for serving powerful interests could be reduced; the administration of scientific funding and publication made more egalitarian; and positive incentives introduced to encourage innovative proposals involving young researchers. Institutionalized protocol and public penalties for malpractice would discourage obvious distortion of research results to favour friends, patrons, political allies, or personal careers.

None of these changes will come easily. The existing systems of science education and administration have evolved to their current form precisely because they fulfil the needs of the established scientists and of their political and industrial benefactors. These powerful interests will energetically oppose the reforms suggested here. But there are indications that other, less powerful but perhaps more broadly representative groups are beginning to recognize the importance of scientific research to their needs as well. Public outrage at current environmental and public-health problems is motivating the formation of organized citizen lobby groups, many of which are intensely interested in the scientific research carried out on their problems. Like other social advocates, these bodies have vested interests in certain research directions and outcomes. Unlike the more powerful industrial and political organizations, however, public-interest groups rarely have the resources or the opportunity to alter the course of scientific investigations. Equal access to such influence is a distant goal, but encouraging initiatives are giving these groups an occasional role in science-policy development.

Environmental groups, for example, have been active in the United States in urging stricter regulation of biotechnology and industrial involvement in academic research,[10] and farmers' coalitions have been lobbying for more broadly beneficial agricultural R&D.[11] Women's groups have changed government attitudes to research on public and occupational health.[12] Less commonly, there are institutionalized arrangements between universities and outside groups: the Université de Québec à Montréal incorporated the idea of public service in its mandate, one example of which is a group called the Action-Research Team on the Biology of Health, a number of researchers in the biology department who collaborate with trade unions in investigating occupational-health problems.[13] There are, therefore, forces working to make scientific education and research more responsive to the public interest, but their activities are still sporadic and narrowly focused.

More widespread, co-ordinated, and politically sophisticated efforts will be needed to extend these individual initiatives to other fields of scientific activity. The current trend to incorporate industrial priorities in scientific-research funding offers a precedent for other groups to demand similar access and influence. The danger, of course, is that the resources of industry and its development and production capabilities are nowhere matched by public-interest groups, which must appeal to the long-term societal benefits and the potential problems avoided by research activity influenced to some extent by social needs.

Improving science education is one of the best ways to alter the expectations of future scientists and their perceptions of how the research process works. Science should be depicted as a personal profession, not a line of work remote from the rest of the community. We must emphasize the importance of human choices in research investigation. We must show how some choices can lead to errors – as, for example, when false assumptions are made regarding the purity of research tools – and how other choices – of research methods or analysis – can steer the findings in certain directions. (Teaching and use of the Method of Multiple Working Hypotheses,[14] as an alternative to the traditional employment of two alternative hypotheses, would help to stress the variety of possible research outcomes and explanations.) Other choices will entirely preclude particular scientific conclusions arising from them: the problems not studied, the hypotheses not tested, or plausible interpretations shunned. In all these ways, we should demonstrate how the identity and aspirations of the scientist have a profound effect on how research work is carried out and what is consequently discovered.

Once the influence of the investigator on the research findings is more universally recognized, active scientists are more likely to devise methods to challenge one another and to counterbalance the inevitable pressures affecting their own work. Professional societies could, for example, expand the portion of their conferences and journals devoted to iconoclastic views and disputed theories. They might experiment with publication formats that permit the exploration of scientific disagreements,[15] or launch new periodicals to consider unconventional or little-known theories or findings.[16]

Politicians, regulatory tribunals, the media, and the general public, all of whom rely on scientific advice when considering regulatory and public-policy issues, would also benefit from a more realistic view of the scientific endeavour. If they understood the impact of vested interests on scientific investigation they would be more likely to recognize the need for advocacy research. They might seek to identify the background and affiliations of scientists giving evidence on public-policy issues, and attempt to contrast this with alternative scientific positions held by researchers with contrary personal inclinations. Minority reports and investigations of the same problem by research teams with different allegiances might be encouraged.

If science education and the publicity surrounding scientific accomplishments took into account the outside pressures on the scientist and the probable effect on the research produced, the inevitability of professional disagreements over scientific evidence would be accepted, if not welcomed. Of course scientific discord does little to help the besieged politicians urgently seeking unequivocal public-policy advice. It would be a constructive lesson to demonstrate that such advice comes only attached to important strings – of bias, values, and assumptions of doubtful validity. This might then force acknowledgement of the non-technical basis for many public-policy decisions and consequently allow greater public, non-expert participation in their formulation. (A technique called "backcasting" has been proposed to help develop socially desirable policies while explicitly acknowledging the assumptions on which they are based.[17] This method of scientific modelling requires the modeller to choose a preferred outcome and to specify the assumptions: the model then generates the range of conditions under which this outcome is possible.)

Changes in our expectations of science and scientists will help us to set wise research priorities and use scientific results more effectively. Such education cannot, however, dismantle the conservative, hierarchical power structure within science that unnecessarily restricts the range of topics, theories, and investigators supported. To alter the current system, which operates to the advantage of established scientists and the fields, theories, and protégés they favour, will require dramatic and, no doubt, unpopular changes in science administration. The necessary reforms fall into the general areas of public disclosure, conflict of interest, peer review, and accountability.

In one area of laboratory science at least, uncertainty and potential errors could be reduced relatively easily. A quality-control program should be established to ensure that the materials, animals, and data produced by commercial supply houses and animal-breeding and -testing laboratories are uncontaminated. The scheme could operate like a traditional insurance policy, whereby policy holders (i.e., the supply houses) would pay a premium to cover compensation in case of complaints of product impurity. Premium charges could be adjusted according to the stringency of laboratory purity protocols, which could be ascertained by random visits and spot checks on the materials or animals supplied.[18]

Government regulations should require this kind of quality-control assurance for all laboratories testing new chemicals in support of registration applications, and for all publicly supported and government research suppliers. If these reforms did not sufficiently improve the rigour of laboratory tests on public-policy matters, such as the public-health effects of new chemicals, then the laboratories or the mechanism for funding them could be taken over by a public agency to maintain a suitable distance between the chemical producer and the chemical tester. Together, these measures would help to standardize the quality of laboratory materials and would raise public confidence in the results of tests supporting new chemical products.

The personal and outside pressures on the academic researcher are much less obvious than those on commercial laboratory scientists. For this reason, public disclosure by academic scientists of all corporate, commercial, governmental, and advocacy-group affiliations should be mandatory. So, too, should past sponsors of each scientist's research projects. If similar disclosure requirements were in effect for journal editors, members of peer-review committees, and those participating in or contributing to regulatory tribunals, it would be much easier consciously to balance the composition of those groups. Identifying and balancing scientists' personal and professional inclinations in this way would help to diversify the research funded and published, and would substantially strengthen public-policy decisions. Not only would regulatory decisions based on scientific evidence likely be fairer, but they would be seen to be fairer.

Public disclosure of scientists' affiliations, past and present, would also discourage private patronage of established researchers in exchange for scientific support for corporate positions; but it would do little to reduce the potential of scientists to derive profit from their own or their students' research efforts funded at public expense. A more comprehensive code of practice is needed to deal with the potential for conflict of interest. Universities and scientific research institutions, journals and funding agencies should all develop conflict-of-interest guidelines that would prohibit individuals with personal financial interests in a field from serving on related peer-group committees and from diverting publicly funded findings to private goals.

In addition, journal and grant-agency guidelines dealing specifically with fraud, plagiarism, republication, and retraction should be developed and circulated to reviewers, who should be encouraged to look out for such deceptions. Those agencies that have not already done so should also develop procedures for reviewing allegations of malpractice and for publicly censuring individuals found to have engaged in these misdemeanours. Professional penalties such as exclusion from professional bodies, exclusion from publishing or from receiving certain grants, and full disclosure of incurred penalties to professional associations will probably be severe enough to deter many would-be frauds, and these measures are certainly cheaper and more quickly accomplished than legal remedies.

The process of developing such policies and guidelines, and their subsequent dissemination, would raise academic consciousness regarding operation of vested interests and would precipitate worthwhile discussion. Enforcement should, as much as possible, be given to ad hoc committees assembled to deal with a specific matter, as further proliferation of ongoing administrative demands on scientists' time could be counterproductive, taking scientists out of their laboratories and adding to the distance between senior scientists and their active research work. Academic, professional, and public watchdogs could also economically ensure compliance with limits on the corporate support for individuals, research projects, or departments, and could guard against the attendant erosion of educational priorities and limitations on academic openness and the free exchange of ideas. In an effort to offset the attraction of commercially useful applied research, a levy on industries benefiting from public support for relevant research could be used to establish a fund for investigations of particular interest to public-advocacy groups.

Reform of the peer-review process must lie at the heart of any effort to disengage science from the vested interests it can serve. Scientific research would be more original, and likely more productive, if the groups deciding who should get funded and published were more broadly based. It is especially important that science-policy development, identification of strategic programs for priority support, and specific funding decisions, which, in effect, control the future direction of scientific investigation, be influenced by the priorities of non-scientists: representatives of the public and of those served by the research in question should be included on grant and science-policy committees.

If there is effective public participation in these decisions there is less need for lay involvement in publication (a much more cumbersome and time-consuming exercise, since so many papers require review and so many journals are highly specialized in content) – the dissemination of whatever findings emerge from the funded studies. Journals' manuscript-review panels, like grant committees, should, however, include many more young scientists at low-status institutions. A deliberate effort should be made to balance the backgrounds of reviewers. For example, in the case of a publication, there might be one specialist reviewer, one reviewer in a related field, and one reviewer suggested by the author, or (possibly less effectively but certainly more efficiently) computer search routines could be used to generate names of suitable reviewers. Journals and granting bodies should seriously consider requiring reviewers to disclose their identities to authors. Applicants and authors should be invited to rate reviewers and they should also be given a chance to reply to carefully explained decisions on their submissions. Special priority should be given, by both journals and grant agencies, to supporting innovative and original work.

Finally, individual scientists must be made more accountable for the work that they do. Many more and smaller grants should be given out, not to laboratory chiefs, but to the often more junior researchers actively involved in the day-to-day operations of the laboratory. Collaboration with other researchers should be encouraged; but the role of each scientist should be clearly defined, and each grant recipient should be responsible for the quality of work carried out with his or her share of the research funds. Journals should restrict co-authorship to those who have made substantial contributions to the research, and the nature of each contribution should be specified in the acknowledgement section of each paper. Grant agencies and journals should require open laboratory procedures and access to data books; university committees should establish research protocols; and granting agencies and professional associations should review charges of infringement of these protocols. In short, the concern of science administrators, funders, and publishers with the important day-to-day operations of research laboratories should be made more explicit, and lapses in laboratory or institutional procedures should become a matter of clear public and scientific interest.

Reforms should not be confined to these measures, nor should they be viewed as rigid or prescriptive. Particular means of addressing the problem of vested interests and scientific research will likely be most usefully developed by those directly involved in each area of research practice and administration. But the single most needed change is one in which we can all participate. The necessary reforms to the administration and practice of scientific research will be precipitated only by a radical revision of the general public's perception of science and scientists, and by recognizing the important role played by the vested interests they serve.

NOTES

Chapter One: Science under Pressure

[1] From Hilary Rose and Steven Rose, *Science and Society* (London: Allen Lane, The Penguin Press, 1969), p. 261.

[2] Ursula Franklin, "Will Women Change Technology or Will Technology Change Women?" in *Knowledge Reconsidered: A Feminist Overview* (Ottawa: Canadian Research Institute for the Advancement of Women, 1984), pp. 77-91.

[3] Bruce V. Lewenstein, "Was There Really a Science Boom?", *Science, Technology and Human Values* 12, issue 2 no. 59 (1987), pp. 29-41.

[4] John Walsh, "Public Attitude Toward Science Is Yes, But – ", *Science* 215 (1982), pp. 270-72.

[5] See, for example, several of David Suzuki's weekly articles in the Saturday issues of the Toronto *Globe and Mail*, from 1987.

[6] William Broad and Nicholas Wade, *Betrayers of the Truth* (New York: Simon and Schuster, 1982).

[7] Paul Brodeur, "Annals of Law", articles published in the *New Yorker* magazine, June 10, 1985, pp. 49-101; June 17, 1985, pp. 45-111; June 24, 1985, pp. 37-77; July 1, 1985, pp. 536-80; and also *An Industry on Trial* (New York: Pantheon Books, forthcoming); Nicholas Hildyard, *Cover Up* (London: New English Library, 1981); Lloyd Tataryn, *Dying for a*

Living (Toronto: Deneau and Greenberg, 1979). See also David Dickson, *The New Politics of Science* (New York: Pantheon Books, 1984), and Ted F. Schrecker, *Political Economy of Environmental Hazards* (a study paper prepared for the Law Reform Commission of Canada, 1984) for more on this subject.

[8] Alan F. Westin, *Whistle Blowing!* (New York: McGraw-Hill, 1981), and L. Freeman, ed., *Nuclear Witnesses* (New York: Norton, 1981). See also "Workshop on Whistle-Blowing in Biomedical Research", September 1981, Agenda Book, available from AAAS Committee on Scientific Freedom and Responsibility, for more on this subject; Brian Martin, "Science Policy: Dissent and Its Difficulties", *Philosophy and Social Action* 12, no. 1 (January-March 1986), pp. 5-23; and B. Martin, C.M.A. Baker, C. Manwell, and C. Pugh, eds., *Intellectual Suppression: Australian Case Histories, Analysis and Responses* (Sydney: Angus and Robertson, 1986).

[9] See, for example, Robert K. Merton, *The Sociology of Science* (Chicago: University of Chicago Press, 1973), p. 270.

[10] Thomas S. Kuhn, *The Structure of Scientific Revolutions* (Chicago: University of Chicago Press, 1962).

[11] See Paul Feyerabend, *Against Method* (London: New Left Books, 1975) and *Science in a Free Society* (London: New Left Books, 1978).

[12] See Richard S. Westfall, "Newton and the Fudge Factor", *Science* 179, (1973) 751-58.

[13] Stephen Jay Gould, "Morton's Ranking of Races by Cranial Capacity", *Science* 200 (1978), pp. 503-09.

[14] Stephen Jay Gould, *The Mismeasure of Man* (New York: W. W. Norton & Co., 1981), p. 66.

[15] Broad and Wade, *Betrayers of the Truth*; C.J. List, "Scientific Fraud: Social Deviance or the Failure of Virtue?" *Science, Technology and Human Values* 10, no. 4 (1985), pp. 27-36.

[16] Harriet Zuckerman, "Deviant Behaviour and Social Control in Science", in Edward Sagarin, ed., *Deviance and Social Change* (Beverly Hills: Sage Publications, 1977), pp. 87-138; H. Zuckerman, "Norms and Deviant Behaviour in Science", *Science, Technology and Human Values* 9, no. 1 (1984), pp. 7-13; Warren Schmaus, "Fraud and the Norms of Human Science", *Science, Technology and Human Values* 8, no. 4 (1983), pp. 12-22.

[17] Broad and Wade, *Betrayers of the Truth*, and Gould, *The Mismeasure of Man*.

[18] Brian Martin, "The Bias of Science", Society for Social Responsibility in Science (ACT), Canberra, Australia; John Bridger Robinson, "Apples and Horned Toads: On the Framework-Determined Nature of the Energy

Debate", *Policy Sciences* 15 (1982), pp. 23-45; Richard Levins and Richard Lewontin, *The Dialectical Biologist* (Cambridge, Mass., and London: Harvard University Press, 1985); and various issues of the *Radical Science Journal*, published by Free Association Books, 26 Freegrove Road, London N7, U.K. Edith Efron offers an ideologically opposite analysis of bias in research on environmental causes of cancer in *The Apocalyptics* (New York: Simon and Schuster, 1984).

[19] See, especially, Michael J. Mahoney, *Scientist as Subject* (Cambridge, Mass.: Bollinger, 1976), and Richard Nisbett and Lee Ross, *Human Inferences* (New Jersey: Prentice-Hall, 1980), as well as Feyerabend, *Against Method* and *Science in a Free Society*. Richard S. Westfall's "Newton and the Fudge Factor" is just one relevant example of the rich field of the history of science. Robert K. Merton has written the classic *The Sociology of Science*, other studies in this field being I. Mitroff's *The Subjective Side of Science* (Amsterdam: Elsevier, 1974), Richard P. Suttmeier's work on Chinese science (for example, "Corruption in Science: The Chinese Case", *Science, Technology and Human Values* 10, no. 1 [1985], pp. 49-61), and Bruno Latour and Steve Woolgar, *Laboratory Life* (London: Sage, 1979). June Goodfield's *An Imagined World* (New York: Penguin, 1981) is a gripping chronicle of one woman's life in science.

[20] John R. Sabine, "The Error Rate in Biological Publication: A Preliminary Survey", *Science, Technology and Human Values* 10, no. 1 (1985), pp. 62-69.

Chapter Two: Science at Risk

[1] Hans Grüneberg, *The Genetics of the Mouse* (The Hague: Martinus Niehoff, 1952), p. 437; Herbert C. Morse, ed., *Origins of Inbred Mice* (New York: Academic Press, 1978).

[2] B. Kahan, R. Auerbach, B.J. Alter, and F.H. Bach, "Histocompatibility and Isoenzyme Differences in Commercially Supplied 'BALB/c' Mice", *Science* 217 (1982), pp. 379-81.

[3] Charles River Laboratories, "A New Generation", publicity material supplied December 1987.

[4] Legal complaint, *Dr. Brenda Kahan* vs. *Charles River Breeding Laboratories Inc.*, filed June 29, 1983, in the U.S. District Court for the Western District of Wisconsin. Case No. 83-C-597 (s), points 31 and 33.

[5] Bruce Gellerman, story editor, *All Things Considered*, National Public Radio, Washington, D.C., July 29, 1983.

[6] Ibid.

[7] *Kahan* vs. *Charles River Breeding Laboratories Inc.*, point 32.

[8] Answer, Ibid., points 25 and 26.

[9] B. Kahan et al., "Histocompatibility", p. 381.

[10] Dr. Clifford Ottaway, University of Toronto, personal communication, 1983.

[11] Dr. Alvin Warfel, personal communication, November 5, 1985.

[12] Ibid.

[13] Ottaway, personal communication, 1983.

[14] Heinrich Bitter-Suermann and M.G. Lewis, Letter to editor: "Rejection of Skin and Spleen 'Isografts' in a Strain of Lewis Rats", *Transplantation* 30 (1980), p. 158.

[15] *Kahan* vs. *Charles River Breeding Laboratories Inc.*

[16] Geoffrey Alman, attorney for Alvin Warfel, personal communication, November 14, 1985.

[17] J.L. Fox, "Scientist Sues Over Genetically Impure Mice", *Science* 221 (1983), p. 626.

[18] David Hanson, lawyer for Kahan, personal communication, November 4, 1985. The Warfel case was settled in January 1986 (G. Alman, personal communication, August 27, 1986).

[19] Gellerman, "All Things Considered".

[20] Editorial: "Responsibility for Trust in Research", *Nature* 289 (1981), p. 212.

[21] W.A. Nelson-Rees and R.R. Flandermeyer, "HeLa Cultures Defined", *Science* 191 (1976), pp. 96-98; W.A. Nelson-Rees, R.R. Flandermeyer, and D.W. Daniels, "Cross-Contamination of Cells in Culture", *Science* 212 (1981), pp. 446-51; W.A. Nelson-Rees and R.R. Flandermeyer, "Inter- and Intraspecies Contamination of Human Breast Tumor Cell Lines HBC and BrCa5 and Other Cell Cultures", *Science* 195 (1977), pp. 1343-44; W.A. Nelson-Rees, R.R. Flandermeyer, and P.K. Hawthorne, "Banded Marker Chromosomes as Indicators of Intraspecies Cellular Contamination", *Science* 184 (1974), pp. 1093-96; W.A. Nelson-Rees, "HeLa Cells and RT4 Cells", *Science* 188 (1975), p. 168; N.L. Harris, D.L. Gang, S.C. Quay, S. Poppema, P.C. Zamecnik, W.A. Nelson-Rees, and S.J. O'Brien, "Contamination of Hodgkin's Disease Cell Cultures", *Nature* 289 (1981), pp. 228-30; W.A. Nelson-Rees, D.W. Daniels, and R.R. Flandermeyer, "Human Embryonic Lung Cells (HEL-R66) Are of Monkey Origin", *Archives of Virology* 67 (1981), pp. 101-4; W.A. Nelson-Rees, R.R. Flandermeyer, and D.W. Daniels, "T-1 Cells Are HeLa and Not of Normal Human Kidney Origin", *Science* 209 (1980), pp. 719-20; W.A. Nelson-Rees, L. Hunter, G.J. Darlington, and S.J. O'Brien,

"Characteristics of HeLa Strains: Permanent vs. Variable Features", *Cytogenetics and Cell Genetics* 27 (1980), pp. 216-31; W.A. Nelson-Rees, V. Klement, W.D. Peterson, Jr., and J.F. Weaver, "Comparative Study of 2 RD, 114 Virus-Indicator Cell Lines, KC and KB", *Journal of the National Cancer Institute* 50 (1973), pp. 1129-35; W.A. Nelson-Rees, V.M. Zhdanov, P.K. Hawthorne, and R.R. Flandermeyer, "HeLa-Like Marker Chromosomes and Type-A Variant Glucose-6-Phosphate-Dehydrogenase Isoenzyme in Human Cell Cultures Producing Mason-Pfizer Monkey Virus-Like Particles", *Journal of the National Cancer Institute* 53 (1974), p. 751; W.A. Nelson-Rees, Letter: "Double Minutes in Human Carcinoma Cell-Lines, with Special Reference to Breast-Tumors: Reply", *Journal of the National Cancer Institute* 63 (1974), p. 537; P.D. Neuwald, C. Anderson, W.O. Salivar, P.H. Aldenderfer, W.C. Dermody, B.D. Weintraub, S.W. Rosen, W.A. Nelson-Rees, and R.W. Rudden, "Expression of Oncodevelopmental Gene-Products by Human Tumor Cells in Culture", *Journal of the National Cancer Institute* 64 (1980), pp. 447-59; W.A. Nelson-Rees, R.R. Flandermeyer, and P.K. Hawthorne, "Distinctive Banded Marker Chromosomes of Human Tumor Cell Lines", *International Journal of Cancer* 16 (1975), pp. 74-82; W.A. Nelson-Rees, "Identification and Monitoring of Cell-Line Specificity", *Progress in Clinical and Biological Research* 26 (1978), pp. 25-79; W.A. Nelson-Rees, "Karyological Considerations and the History of Cell Culture", *In Vitro Monograph* no. 5, pp. 142-51; "Uses and Standardization of Vertebrate Cell Cultures", Tissue Culture Association, 1984. The HeLa problem has been explored in the popular book by Michael Gold, *A Conspiracy of Cells: One Woman's Immortal Legacy and the Medical Scandal It Caused* (Albany, N.Y.: State University of New York Press, 1986).

[22] W.A. Nelson-Rees, D.W. Daniels, and R.R. Flandermeyer, "Cross-Contamination of Cells in Culture", *Science* 212 (1981), p. 451.

[23] Interestingly, one of the first cases of cell-culture line contamination reported publicly was discovered in Toronto by, among others, Louis Siminovitch, who figures in Chapter Five (K.H. Rothfels, A.A. Axelrod, L. Siminovitch, E.A. McCulloch, and R.C. Parker, "The Origin of Altered Cell Lines from Mouse, Monkey, and Man, as Indicated by Chromosome and Transplantation Studies", *Proceedings of the Canadian Cancer Research Conference* 3 [1959], pp. 189-214).

[24] Z. Jaworowsky, M. Bysiek, and L. Kownacka, "Flow of Metals into the Global Atmosphere", *Geochimica et Cosmochimica Acta* 45 (1981), p. 2185.

[25] Ibid., p. 2196.

[26] C.C. Patterson, "Criticism of 'Flow of Metals into the Global Atmosphere,'" *Geochimica et Cosmochimica Acta* 47 (1983), p. 1163.

[27] C.C. Patterson, "Analyses for Lead at Levels of 10^{-13} to 10^{-9} g/g", unpublished speech to U.S. Food and Drug Association chemists in 1982.

[28] J. Nriagu, Canadian Centre for Inland Waters, Burlington, Ont., personal communication, October 14, 1987.

[29] W.L. Athey, "Letter to Editor, re Rejection of Islet 'Isografts' in a Strain of Lewis Rats", *Transplantation* 31 (1981), p. 231.

[30] Herbert C. Morse, personal communication, 1983.

[31] M.F. Festing, "Genetic Reliability of Commercially-Bred Laboratory Mice", *Laboratory Animals* 8 (1974), pp. 265-70.

[32] For a discussion of published errors and their implications, see John R. Sabine, "The Error Rate in Biological Publications: A Preliminary Survey", *Science, Technology and Human Values* 10, no. 1 (1985), pp. 62-69.

Chapter Three: Scientists Fool Themselves

[1] Robert J. Gullis, "Statement", *Nature* 265 (1977), p. 764, as reported in Warren Schmaus, "Fraud and the Norms of Science", *Science, Technology and Human Values* 8, no. 4 (1983), p. 13.

[2] William C. Clark and Giandomenico Majone, "The Critical Appraisal of Scientific Inquiries with Policy Implications", *Science, Technology and Human Values* 10, no. 3 (1985), p. 8.

[3] The material presented on this story relies heavily on information derived from: A.G. Levine, *Love Canal: Science, Politics, and People* (Lexington, Mass.: Lexington Books, 1982); B. Paigen, "Controversy at Love Canal", *The Hastings Center Report* 12, no. 3 (June 1982), pp. 29-37.

[4] R. Whalen, Order: State of New York: Department of Health in the Matter of the Love Canal Chemical Waste Landfill Sites Located in the City of Niagara Falls. Albany, New York, August 2, 1978, as cited in Levine, *Love Canal*, p. 59.

[5] D. Shribman and P. MacClennan, "Wider Patterns of Canal Illness Found: Stream Beds Trace Paths of Sickness", *Buffalo Evening News*, October 18, 1978, as cited in Levine, *Love Canal*, p. 93.

[6] P. MacClennan, "State Extends Probes North of Canal", *Buffalo Evening News*, January 5, 1979, as cited in Levine, *Love Canal*, p. 93.

[7] N.J. Vianna, A.K. Polan, R. Regal, S. Kim, G.E. Haughie, and D. Mitchell, *Adverse Pregnancy Outcomes in the Love Canal Area*, Provisional Report for the New York State Department of Health, April 1980.

[8] Ibid., pp. 20-21.

[9] "Love Canal, A Special Report to the Governor and Legislature of New York State", April 1981, by New York State Department of Health, p. 44; Karen Kalaijain, New York State Department of Health, personal communication October 28, 1987, confirms that ninety-six toxic waste sites in Niagara County are now known.

[10] Beverley Paigen, personal communication, August 21, 1984.

[11] Ibid.

[12] Nicholas Ashford, in "Examining the Role of Science in the Regulatory Process", *Environment* 25, no. 5 (1983), p. 7.

[13] Nicholas J. Vianna, New York Department of Public Health, personal communication, August 28, 1984.

[14] Nicholas J. Vianna and Adele K. Polan, "The Incidence of Low Birth Weight Among Love Canal Residents", *Science* 226 (1984), p. 1217.

[15] Harvey Brooks, "The Resolution of Technically Intensive Public Policy Disputes", *Science, Technology and Human Values* 9, no. 1 (1984), pp. 39-50.

[16] R.L. Meehan, *The Atom and the Fault* (Cambridge, Mass.: MIT Press, 1984).

[17] Royal Society of Canada, Commission on Lead in the Environment, "Lead in the Canadian Environment", September 1986, and Appendix XI, dissent.

[18] Nick Fillmore and Anne Wordsworth, "The Blast Supper", *This Magazine* 21, no. 1 (March-April 1987), pp. 14-22.

[19] Clark and Majone, "Critical Appraisal of Scientific Inquiries".

[20] See also letters to *The Lancet* from R. Stephens and R. Russell-Jones, October 27, 1984, p. 976, and from P.C. Elwood, A. Essex-Cater, and R.C. Robb, August 11, 1984, p. 355.

[21] C. Duncan, R.A. Kusiak, J. O'Heany, L.F. Smith, L. Spielberg, and J. Smith, "Blood Lead and Associated Risk Factors in Ontario Children, 1984: Summary and Conclusions", (Ontario Ministries of Environment, Health, and Labour, 1985), p. 20.

[22] James L. Pirkle, personal communication, September 6, 1984.

[23] Vernon Houk, Director of the Center for Environmental Health, Centers for Disease Control, Atlanta, Ga., U.S.A., personal communication to K. Cooper, October 8, 1987.

[24] James L. Pirkle, in a conversation with Barbara Wallace, December 6, 1985.

[25] Royal Society of Canada, Commission on Lead in the Environment, "Lead in Gasoline", September 1985, p. 55.

[26] Centers for Disease Control, "Blood-Lead Levels in U.S. Population", *Morbidity and Mortality Weekly Report* 31, no. 10 (March 19, 1982), pp. 132-34.

[27] J.L. Pirkle, "Chronological Trend in Blood Lead Levels in the United

States Between 1976 and 1980", unpublished paper presented at the International Conference on Heavy Metals in the Environment, Heidelberg, September 1983.

[28] J. Schwartz, "The Relation Between Gas Lead and Blood Lead in Americans", unpublished paper presented at the International Conference on Heavy Metals in the Environment, Heidelberg, September 1983.

[29] D.R. Lynam, G.A. Hughmark, B.F. Fort, Jr., and C.A. Hall, "Blood Lead Concentrations and Gasoline Lead Usage", *Heavy Metals in the Environment, International Conference: Heidelberg, September 1983,* vol. 1, p. 417.

[30] J.M. Pierrard, C.G. Pfeifer, and R.D. Snee, "Assessment of Blood Lead Levels in the U.S.A. from NHANES II Data", *International Conference on Heavy Metals in the Environment: Heidelberg, September 1983,* vol. 1, p. 421.

[31] Pirkle, "Chronological Trend in Blood Lead Levels".

[32] J.L. Pirkle, personal communication, September 6, 1984.

[33] Lynam et al., "Blood Lead Concentrations"; Pierrard et al., "Assessment of Blood Lead Levels in the U.S.A."

[34] Pierrard et al., ibid.; see also "Analysis of Reported Blood Lead Level Declines During NHANES II and Their Relationship to the Survey Design and Gasoline Lead Exposure", E.I. du Pont de Nemours and Co., Inc. (DuPont), February 14, 1984.

[35] J. Rosenblatt, H. Smith, R. Royall, R. Little, and J.R. Landis, *Report of the NHANES II Time Trend Analysis Review Group* (North Carolina: United States Environmental Protection Agency, June 15, 1983), pp. 10 and 11.

[36] Lynam et al., "Blood Lead Concentrations", p. 420.

[37] Rosenblatt et al., *Report,* pp. 12 and 14.

[38] United States Environmental Protection Agency, "EPA Proposes Controls on Lead in Gasoline", *Environmental News,* July 30, 1984.

[39] W.D. Ruckelshaus (administrator of USEPA), "Press Conference Announcing Proposed Reductions of Lead in Gasoline", Washington, D.C., July 30, 1984.

[40] F.D. Gottwald, Jr., and B. C. Gottwald, *Ethyl Corporation 1983 Annual Report* (Richmond, Va.: Ethyl Corporation, 1983), p. 3.

[41] D.R. Lynam, personal communication, August 22, 1984.

[42] D.C.G. Pfeifer, personal communication, August 1984.

[43] Gottwald and Gottwald, *Annual Report,* p. 18.

[44] D.R. Lynam, personal communication, August 22, 1984; D.C.G. Pfeifer, personal communication, August 1984.

[45] E. Efron, *The Apocalyptics* (New York: Simon and Schuster, 1984).

[46] D.R. Royall, personal communication, October 20, 1984.

[47] See, for example, Constance Holden, "Scientist with Unpopular Data

Loses Job", *Science* 21 (1980), pp. 749-50; and Alan F. Weston, ed., *Whistle-Blowing* (New York: McGraw-Hill Books, 1981).

[48] June Goodfield, *An Imagined World* (New York: Harper & Row Publishers, 1981).

[49] Evelyn Fox Keller, *A Feeling for the Organism* (New York: W.H. Freeman, 1983).

[50] Robert Rosenthal and Donald B. Rubin, "Interpersonal Expectancy Effects: The First 345 Studies", *The Behavioral and Brain Sciences* 3 (1978), pp. 377-415.

[51] C.G. Lord, L. Ross, and M.R. Lepper, "Biased Assimilation and Attitude Polarization: The Effects of Prior Theories on Subsequently Considered Evidence", *Journal of Personality and Social Psychology* 37, no. 11 (1979), pp. 2098-2109, as referenced in Richard E. Nisbett and Lee Ross, *Human Inference: Strategies and Shortcomings of Social Judgement* (New York: Prentice-Hall, 1980), p. 170.

[52] Thomas S. Kuhn, "Normal Measurement and Reasonable Agreement", in *Science in Context*, ed. Barry Barnes and David Edge (Milton Keynes, U.K.: Open University Press, 1982), p. 89.

[53] Brian Wynne, "Natural Knowledge and Social Context: Cambridge Physicists and the Luminiferous Ether", in Barnes and Edge, *Science in Context*, p. 228.

[54] P.B. Medawar, *Advice to a Young Scientist* (New York: Harper & Row Publishers, 1979), p. 6.

[55] Donald T. Campbell, "Assessing the Impact of Planned Social Change", *Evaluation and Program Planning* 2 (1979), pp. 67-69.

[56] Thomas S. Kuhn, *The Structure of Scientific Revolutions* (Chicago: University of Chicago Press, 1962).

Chapter Four: Public Knowledge and Private Interests

[1] "Lawyer Asked to Probe Biology Controversy", *Globe and Mail*, January 10, 1984, p. 9; D. Wimhurst, "McGill Appoints Lawyer to Investigate Two Professors' Company Activities", *The Gazette*, Montreal, January 10, 1984, p. A-5.

[2] D. Wimhurst, "Professor Leaves McGill for Year as Controversy Boils Around Firm", *The Gazette*, Montreal, November 15, 1983, p. A-6; D. Wimhurst, "McGill Professors in Ethics Dispute – Department Is Split on Whether Private Lab Should Be on Campus", *The Gazette*, Montreal, October 29, 1983, p. B5;   A.K. Paterson, "A Report to the Principal of

McGill University Concerning Events in the Department of Microbiology and Immunology from 1981 to 1983, Including Recommendations with respect to the Involvement of Academic Personnel in Commercial Activities", March 1984, pp. 82-83.

[3] Paterson, "A Report to the Principal", p. 25.

[4] Sheldon Krimsky, "The Capture of Genetic Technologies", *Science for the People* 17, no. 3 (1985), p. 36.

[5] W. Boly, "The Gene Merchants", *California*, September 1982, p. 79; B.J. Culliton, "The Hoechst Department at Mass General", *Science* 216 (1982), pp. 1200-03; David E. Sanger, "Corporate Links Worry Scholar", *Sunday New York Times*, October 17, 1982.

[6] Dorothy Nelkin and Richard Nelsen, with Casey Kierman, "Commentary: Academic-Industrial Alliances", *Science, Technology and Human Values* 12, no. 1 (Winter 1987), pp. 65-74.

[7] Boly, "The Gene Merchants", p. 79.

[8] Colin Norman, "Electronics Firms Plug into the Universities", *Science* 217, (1982), p. 514.

[9] B.J. Culliton, "Harvard and Monsanto: The $23-Million Alliance", *Science* 195, (1977), pp. 759-63.

[10] Glenn Wheeler, "Scoring Research Revenue", *Now* (Toronto), July 9-15, 1987, pp. 9-10.

[11] R. Matas, "Big Companies, Universities Forging New Link", *Globe and Mail*, May 4, 1984, p. M4.

[12] University of Toronto *Bulletin*, October 26, 1987, p. 13.

[13] J. Miller, "The New Paradise and the Potential Fall: Biotechnology and the University Science Community", in *Science in Society: Its Freedom and Regulation*, eds. Fraser Homes-Dixon and Anne T. Perkins, (Ottawa: CSP Publications, 1982), p. 140.

[14] For example, the University of Toronto has assembled a Research Ancillaries Advisory Group, which reported in November 1982, and the University of Guelph has published an Industrial Interaction Strategy to promote industry/university collaboration.

[15] S. Strauss, "Cuts Will Take Place, Science Council Told", *Globe and Mail*, July 26, 1985, p. 5.

[16] Allen Good, "Private Research a Doubtful Answer", Toronto *Globe and Mail*, April 23, 1987, p. A7; Donald C. Savage, CAUT report on university research, *CAUT Bulletin*, September 1987, pp. 11-17.

[17] *Advisory Board for the Research Councils*, Report of the Working Party on the Private-Sector Funding of Scientific Research, published May 1986, as quoted in *The Guardian*, May 21, 1986, p. 4, in an article by Sarah Boseley.

[18] Colin Norman, "Industrial R&D Rises", *Science* 217 (1982), p. 427.

[19] B.J. Culliton, "Monsanto Gives Washington U.S.$23.5 Million", *Science* 216, (1982), pp. 1295-96.

[20] For an exploration of many of these problems, see *The Science Business: Report of the Twentieth-Century Fund Task Force on the Commercialization of Scientific Research, with a background paper by Nicholas Wade* (New York: Priority Press, 1984); and also *Science, Technology and Human Values* 12, no. 1 (Winter 1987), the issue devoted to the "Private Appropriation of Public Research."

[21] J. Eisen, "The Academic-Industrial Complex", transcript of CBC *Ideas* radio series aired November 1982, pp. 13-14.

[22] Pamela Samuelson, "Innovations and Competition: Conflicts Over Intellectual Property Rights in New Technologies", *Science, Technology and Human Values* 12, no. 1, p. 6-21.

[23] David Baltimore, in "Discussion", in *Genetics and the Law III*, eds. Aubrey Milunsky and George J. Annas (New York and London: Plenum Press, 1985), p. 65.

[24] A.H. Meyerhoff, "The United States House of Representatives Hearings on University/Industrial Co-operation in Biotechnology", presented to the Committee on Science and Technology, June 16, 1982. See also Milunsky and Annas, eds., *Genetics and the Law III*.

[25] L. Wofsy, "The Academic-Industrial Complex", transcript of CBC *Ideas* radio series produced by J. Eisen and aired November 1982, p. 12.

[26] Nelkin and Nelson, "Academic-Industrial Alliances", p. 72.

[27] B. Silcock, "The Lab That Makes Geniuses", *The Sunday Times*, London, October 24, 1982, p. 17.

[28] Meyerhoff, "Hearings on University/Industrial Co-operation", pp. 3-4.

[29] B.J. Culliton, "The Academic-Industrial Complex", *Science* 216 (1982), pp. 960-62, and Nelkin and Nelson, "Academic-Industrial Alliances", p. 72.

[30] J.W. Servos, "The Industrial Relations of Science: Chemical Engineering at MIT, 1900-1939", ISIS 71, no. 259 (1980), pp. 531-49.

[31] Boly, "The Gene Merchants", p. 176.

[32] Meyerhoff, "Hearings on University/Industrial Co-operation".

[33] McGill Ad Hoc Senate Committee on Proprietary Research and Academic Freedom, Guidelines with Respect to Research and Similar Activities, October 1984.

[34] Ralph Abascal, personal communication, October 21, 1987.

[35] Ibid.

[36] Eisen, "The Academic-Industrial Complex", p. 4.

[37] Boly, "The Gene Merchants", p. 176.

[38] Peter Von Stackelberg, personal communication, 1982.

[39] Kevin Cox, Toronto *Globe and Mail*, June 30, 1983, p. A17.

[40] N.A. Ashford, "A Framework for Examining the Effects of Industrial Funding on Academic Freedom and the Integrity of the University", *Science, Technology and Human Values* 8, Issue 2 (Spring 1983), pp. 21-22.

[41] L. Marks, London Observer Service, "Tenure System Under Fire at British Universities", Toronto *Globe and Mail*, August 7, 1987, p. A7.

[42] Servos, "The Industrial Relations of Science", p. 546.

[43] B.M. Owen and R. Braeutigam, *The Regulation Game* (Cambridge, Mass.: Bollinger, 1978), p. 7.

[44] Ad Hoc Senate Committee on Proprietary Research and Academic Freedom, McGill University, "Guidelines with Respect to Research and Similar Activities", October 1984, and Association of American Universities, "University Policies in Conflict of Interest and Delay of Publication", February 1985.

[45] L. Wofsy, "Biology and the University on the Market Place: What's for sale?", transcript of a lecture given March 16, 1982, at the University of California at Berkeley, pp. 16-17.

[46] D. Noble, "Academia Incorporated", *Science for the People* 15, no. 1 (Jan./Feb. 1983), pp. 7-11, 50-52; R.H. Linnell, ed., Dollars and Scholars (Los Angeles: University of Southern California, 1982).

[47] This is advocated by Dr. Arnold Relman, editor of *The New England Journal of Medicine*, in Milunsky and Annas, eds., *Genetics and the Law III* (1985), p. 67.

[48] Gabriele Bammer, Ken Green, and Brian Martin, "Who Gets Kicks out of Science Policy?" *Search* 7, nos. 1-2 (1986), pp. 41-46.

[49] Twentieth-Century Fund Task Force, *The Science Business*, p. 76.

Chapter Five: The Academic Hustle

[1] Charles Babbage, *Reflections on the Decline of Science in England* (London: n.p., 1830), pp. 174-83.

[2] Allan Franklin, "Forging, Cooking, Trimming and Riding on the Bandwagon", *American Journal of Physics* 52, no. 9 (1984), pp. 786-93; William Broad and Nicholas Wade outline many historical cases of scientific misconduct in their book *Betrayers of the Truth* (New York: Simon and Schuster, 1982).

[3] O.W. McBride and R.S. Athwal, "Chromosome-Mediated Gene Trans-

fer with Resultant Expression and Integration of the Transferred Genes in Eukaryotic Cells", *Brookhaven Symposium on Biology* 29 (1978), pp. 119-20.

[4] Taped interview with Dr. L. Siminovitch, January 1982.

[5] See D.A. Spandidos, and L. Siminovitch, "Transfer of Codominant Markers by Isolated Metaphase Chromosomes in Chinese Hamster Ovary Cells" [1977a], *Proceedings of the National Academy of Sciences USA* 74, (1977), p. 3480-84; "Linkage of Markers Controlling Consecutive Biochemical Steps in CHO Cells as Demonstrated by Chromosome Transfer" [1977b], *Cell* 12 (1977), p. 235-42; "Transfer of Anchorage Independence by Isolated Metaphase Chromosomes in Hamster Cells" [1977c], *Cell* 12 (1977), pp. 675-82; "Genetic Analysis by Chromosome-Mediated Gene Transfer in Hamster Cells" [1977d], *Brookhaven Symposium on Biology* 29 (1977), p. 127-34; "Transfer of the Marker for Morphologically Transformed Phenotype by Isolated Metaphase Chromosomes in Hamster Cells" [1978a], *Nature* 271 (1978), p. 259-61; "The Relationship Between Transformation and Somatic Mutation in Human and Chinese Hamster Cells" [1978b], *Cell* 13 (1978), pp. 651-62.

[6] Siminovitch, taped interview.

[7] Bill Lewis, personal communication, 1981; Brian D. Johnson, "Finding Foul Play in the Laboratory", *Maclean's* 95, no. 1 (January 4, 1982), pp. 41-44.

[8] W.H. Lewis, P.R. Srinivasan, N. Stokoe, and L. Siminovitch, "Parameters Governing the Transfer of the Genes for Thymidine Kinase and Dihydrofolate Reductase into Mouse Cells Using Metaphase Chromosomes or DNA", *Somatic Cell Genetics* 6 no. 3 (1980), p. 334.

[9] Siminovitch, taped interview.

[10] Demetrios Spandidos, in an interview with Graham Beakhust, October 1981.

[11] Form letter from D. Spandidos, Athens, June 1978.

[12] Science Citation Index 1985, vol. 7; Demetrios A. Spandidos and Neil M. Wilkes, "Malignant Transformation of Early Passage Rodent Cells by a Single Mutated Human Oncogene", *Nature* 310 (1984), p. 469.

[13] Johnson, "Finding Foul Play in the Laboratory"; Broad and Wade, "Betrayers of the Truth"; Patricia Woolf, "Fraud in Science: How Much, How Serious?" *The Hastings Centre Report* 11, no. 5 (October 1981), pp. 9-14; National Institutes of Health, memorandum from Associate Director for Extramural Research and Training to Director, re Decision on Alleged Misconduct at Brigham and Women's Hospital, January 5, 1983.

[14] Claudio Milanese, Neil E. Richardson, Ellis L. Reinherz, "Retraction of Data", *Science* 234 (1986), p. 1056.

[15] "Medical Scientist Censured for Misrepresenting Data", Toronto *Globe*

and Mail, July 18, 1987, p. A6; National Institutes of Health, Office of Extramural Research and Training, Freedom of Information package on the Glueck case, July 1987, p. 14.

[16] Philip M. Boffey, "U.S. Study Finds Fraud in Top Researcher's Work on Mentally Retarded", *New York Times National News*, May 24, 1987, p. 12Y; various documents and memoranda, National Institute of Mental Health, May 8, 1987.

[17] Edward T. Brandt, Jr., "PHS Perspectives on Misconduct in Science", *Public Health Reports* 98 (1983), pp. 136-39; George Howe Colt, "Too Good to Be True", *Harvard Magazine* (1983), pp. 22-28; Mary Miers, personal communication to Daryl Chubin, May 1984, all as reported in Daryl Chubin, "Misconduct in Research", *Minerva* 23, no. 2 (1985), p. 178.

[18] I. St. James-Roberts, "Cheating in Science", *New Scientist* 72, no. 1028 (November 25, 1976), pp. 466-69.

[19] Philip M. Boffey, "Rise in Science Fraud Seen; Need to Win Cited as a Cause", *New York Times*, May 30, 1985.

[20] Daniel E. Koshland, Jr., "Fraud in Science", *Science* 235, no. 4785 (1987), p. 141.

[21] G.B. Kolata, "Reevaluation of Cancer Data Eagerly Awaited", *Science* 214 (1981), pp. 316-18; P. Newmark, "Hidden Spectre", *Nature* 293 (1981), p. 329.

[22] E. Racker and M. Spector, "Warburg Effect Revisited: Merger of Biochemistry and Molecular Biology", *Science* 213 (1981), p. 303.

[23] E. Racker, Letter: "Warburg Effect Revisited", *Science* 213 (1981), p. 1313.

[24] K. McKean, "A Scandal in the Laboratory", *Discover* (November 1981), pp. 18-23; N. Wade, "The Rise and Fall of a Scientific Superstar", *New Scientist* (September 24, 1981), pp. 781-82.

[25] Quote from Report of the International Commission of Inquiry into the Scientific Activities of Professor Karl Illmensee, January 30, 1984, p. 27; Report of the Ad Hoc Committee to Investigate the Collaboration of Drs. Illmensee and Hoppe, June 2, 1983; J. McGrath and D. Solter, "Nuclear Transplantation in the Mouse Embryo by Microsurgery and Cell Fusion", *Science* 220 (1983), pp. 1300-02; Staff of *Nature*, "Coat Clue to How Mice Can Be Cloned", *The Times* (London), June 20, 1983, p. 2; E.P. McQuaid, "Scientist Faces Clone Allegations", *Times Higher Education Supplement*, June 10, 1983, p. 7.

[26] B. Lewin, personal communication, 1982.

[27] Racker, "Warburg Effect Revisited".

[28] V.M. Vogt, R.B. Pepinsky, and E. Racker, "SRC Protein and the Kenase Cascade", *Cell* 25 (September 1981), p. 827.

[29] Spandidos and Siminovitch, op. cit. (1977a,b,c,d, 1978a).

[30] Johnson, "Finding Foul Play in the Laboratory".

[31] Siminovitch, taped interview.

[32] Lewis, Srinivasan, Stokoe, and Siminovitch, "Parameters Governing the Transfer of the Genes for Thymidine Kinase and Dihydrofolate Reductase into Mouse Cells Using Metaphase Chromosomes or DNA", p. 334.

[33] Siminovitch, taped interview.

[34] Science Citation Index 1980-84, vol. 26; 1985, vol. 7; 1986, vol. 8; 1987.

[35] O.W. McBride and J.L. Peterson, "Chromosome Mediated Gene Transfer in Mammalian Cells", *Annual Review of Genetics* 14 (1980), pp. 321-45.

[36] Assorted memoranda provided by the NIH concerning the reviews of research by Dr. Pieter Kark, at the University of California, Los Angeles, Dr. Marc J. Straus at Boston University Medical Centre, Dr. Arthur H. Hale at Bowman Gray School of Medicine, and Dr. Charles J. Glueck at the University of Cincinnati College of Medicine.

[37] Various documents and memoranda, National Institute of Mental Health, May 8, 1987.

[38] William J. Broad, "Imbroglio at Yale (I): Emergence of a Fraud", *Science* 210 (1980), p. 39, and "Imbroglio at Yale (II): A Top Job Lost", *Science* 210 (1980), pp. 171-73; Morton Hunt, "A Fraud That Shook the World of Science", *New York Times Magazine* (November 1981), pp. 42-58, 68-75.

[39] B.J. Culliton, "Coping with Fraud: The Darsee Case", *Science* 220 (1983), p. 31.

[40] Broad and Wade, "Betrayers of the Truth".

[41] James M. Roxburgh, personal communication, January 1982.

[42] Ibid.

[43] Siminovitch, taped interview.

[44] Robert Weinberg, personal communication, January 26, 1982.

[45] See, for example, Peter Newmark, "Disputed X-ray Data Unresolved", *Nature* 303 (1983), p. 197; Claudio M. Milanese et al., *Science* 234 (1986), p. 1056; Philip M. Boffey, "U.S. Study Finds Fraud in Top Researchers Work on Mentally Retarded", *New York Times*, May 24, 1987, p. 2Y; "Medical Scientist Censured for Misrepresenting Data", Toronto *Globe and Mail*, July 8, 1987, p. A6.

[46] Broad, "Imbroglio at Yale".

[47] Siminovitch, taped interview.

[48] Colin Norman, "The Life and Times of an Academic Scientist", *Science* 214 (1981), p. 37.

[49] Gairdner Foundation citation, 1981, p. 8.

[50] Izaak Walton Killam Memorial Prizes Brochure, 1981.

[51] Siminovitch, taped interview.

[52] Ibid.

[53] B. Lewin, personal communication, 1982.

[54] Walter W. Stewart and Ned Feder, "The Integrity of the Scientific Literature", *Nature* 325 (1987), pp. 207-14.

[55] Ibid., p. 207.

[56] Theodore D. Sterling, "Publication Decisions and Their Possible Effects on Inferences Drawn from Tests of Significance – Or Vice Versa", *American Statistical Association Journal* 54 (March 1959), p. 30.

[57] Ibid.

[58] William J. Broad, "Yale Adopts Plan to Handle Charges of Fraud", *Science* 217 (1981), p. 311; David Dickson, "Data Falsification: Harvard Acts", *Nature* 294 (1981), p. 684; Mary L. Miers, Institutional Liaison Officer, Office of Extramural Research and Training, National Institutes of Health, U.S.A., personal communication, November 29, 1983; Penelope J. Greene, Jane S. Durch, Wendy Horowitz, and Valwyn S. Hooper, "Policies for Responding to Allegations of Fraud in Research", *Minerva* 23, no. 2 (1985), pp. 203-15. See also Association of American Medical Colleges, "The Maintenance of High Ethical Standards in the Conduct of Research" (Washington, D.C., 1982); and Report of the Association of American Universities Committee on the Integrity of Research; U.S. Public Health Service, "Policies for Handling Misconduct in Science", in Edward N. Brandt, Jr., "PHS Perspectives on Misconduct in Science", *Public Health Reports* 98, no. 2 (1983), p. 139; and C. Ian Jackson and John W. Prados, "Honor in Science", *American Scientist* 71 (September/October 1983), pp. 462-64.

[59] Daryl E. Chubin, "Research Malpractice", draft manuscript of paper in *BioScience* (February 1985).

[60] Eugene Braunwald, "On Analysing Scientific Fraud", *Nature* 325 (1987), p. 216.

[61] Daryl E. Chubin, "Misconduct in Research", *Minerva* 23, no. 2 (1985), pp. 175-202; Braunwald, "On Analysing Scientific Fraud".

Chapter Six: Getting into Print

[1] Morris Millner, oral communication in 1982 talk to Canadian Medical and Biological Engineering Society, Fredericton, New Brunswick

[2] L. Gladysz, personal communication from *The New England Journal of Medicine* editorial assistant, January 1986.

[3] R. Campbell, letter: "Almighty Referee", *New Scientist* 91, no. 1260 (1981), p. 42 (with data first reported by J. McCartney at the First International Conference of Scientific Editors, 1977).

[4] D.P. Peters and S.J. Ceci, "Peer-review Practices of Psychological Journals: The Fate of Published Articles Submitted Again", *The Behavioral and Brain Sciences* 5, no. 2 (June 1982), pp. 187-255.

[5] D. Lazarus, "Interreferee Agreement and Acceptance Rates in Physics", *The Behavioural and Brain Sciences* 5, no. 2 (June 1982), p. 219.

[6] For example, *The Proceedings of the National Academy of Sciences* publishes only articles of four pages or less (R. Sheinin, personal communication, March 9, 1986).

[7] W.J. Broad, "The Publishing Game: Getting More for Less", *Science* 211 (1981), pp. 1137-39.

[8] Ibid.

[9] Philip M. Boffey, "U.S. Study Finds Fraud in Top Researcher's Work on Mentally Retarded", *New York Times National News*, May 24, 1987, p. 12Y; various documents and memoranda, National Institute of Mental Health, May 8, 1987.

[10] National Institute of Mental Health, Final Report, Investigation of Alleged Scientific Misconduct, pp. 42-43.

[11] Walter W. Stewart and Ned Feder, "The Integrity of the Scientific Literature", *Nature* 325 (January 15, 1987), pp. 207-14.

[12] Broad, "The Publishing Game".

[13] C.J. Sindermann, *Winning the Games Scientists Play* (New York: Plenum Press, 1982), p. 20.

[14] Correspondence, "Duplicate Publication on Postmenopausal Bone Loss", *The New England Journal of Medicine* 317, no. 13 (September 24, 1987), pp. 833-34.

[15] J.M. Lauweryns, J. Baert, and W. De Loecker, "Fine Filaments in Lymphatic Endothelial Cells", *The Journal of Cell Biology* 68 (1976), pp. 163-67; and "Intracytoplasmic Filaments in Pulmonary Lymphatic Endothelial Cells – Fine Structure and Reaction After Heavy-Meromyosin Incubation", *Cell and Tissue Research* 163, no. 2 (1975), pp. 111-24.

[16] Joseph M. Lauweryns and Marnix Cokelaere, "Hypoxie-sensitive Neuro-epithelial Bodies Intrapulmonary Secretory Neuroreceptors Modulated by the CNS", *Zeitschrift für Zellforschung und Mikroskopische Anatomie* 145 (1973), pp. 521-40; and "Intrapulmonary Neuro-Epithelial Bodies: Hypoxia-Sensitive Neuro(Chemo-)Receptors", *Experientia* 29, no. 11 (1973), pp. 1384-86.

[17] Figure 4 in K. Izutsu, "Irradiation of Parts of Single Mitotic Apparatus in Grasshopper Spermatocytes with an Ultraviolet Microbeam", *Mie*

Medical Journal 9 (1959), pp. 15-29, is identical to Figure 15 in K. Izutsu, "Effects of Ultraviolet Microbeam Irradiation upon Division in Grasshopper Spermatocytes. II. Results of Irradiation During Metaphase and Anaphase I", *Mie Medical Journal* 11 (1961), pp. 213-32; and Figure 14 of Izutsu (1961) is identical to Figure 15 of S. Takeda and K. Izutsu, "Partial Irradiation of Individual Mitotic Cells with Ultraviolet Microbeam", *Symposia for Cellular Chemistry (Saibo Seibutfugaku Shimpojiumu)* 10 (1960), pp. 245-62; as reported in P.J. Sillers, D. Wise, and A. Forer, "Prometaphase Forces Towards Opposite Spindle Poles Are Not Independent: An On/Off Control System Is Identified by Ultraviolet Microbeam Irradiations", *Journal of Cell Science* 64 (1983), p. 84.

[18] Hidemi Sato, "Role of Spindle Micro-Tubules for the Anaphase Chromosome Movements in Fertilized Sea Urchin Eggs", *Cell Differentiation* 11 (1982), pp. 345-48; and H. Sato, T. Kato, T.C. Takahashi, and T. Ito, "Analysis of D_2O Effect on *in vivo* and *in vitro* Tubulin, Polymerization and Depolymerization", Chapter 20 in H. Sakai, H. Mohri, and G.G. Borisy, eds., *Biological Functions of Microtubules and Related Structures* (New York: Academic Press, 1982).

[19] P.M. Boffey, "Journals Combat Scientists' Deceit in Submitting Study Reports", *New York Times*, December 14, 1982, p. 17; and N. Wade, "Medical Journal Draws Lancet on Rival", *Science* 211 (1981), p. 561.

[20] P.H. Abelson, Editorial: "Excessive Zeal to Publish", *Science* 218, (1982), p. 953.

[21] A.S. Relman, "The Patient and the Press", *Bryn Mawr Alumnae Bulletin,* Fall 1981, pp. 2-5.

[22] Brian Martin discusses the varieties of plagiarism and suggests some measures to curtail it in "Plagiarism and Responsibility", *Journal of Tertiary Educational Administration* 6, no. 2 (October 1984).

[23] D.E. Chubin, "Research Misconduct: An Issue of Science Policy and Practice", Final Project Report to the Ethics and Values in Science and Technology Program of the United States National Science Foundation, 1984.

[24] W.J. Broad, "Would-Be Academician Pirates Papers", *Science* 208 (1980), pp. 1438-40 (quote from p. 1440); and W.J. Broad and N. Wade, *Betrayers of the Truth* (New York: Simon and Schuster, 1982); part of the paragraph of text ending with reference 24 is taken from my text for CBC Radio *Ideas* series, "Science and Deception", December 1982, Part IV, p. 1.

[25] A. von Wiess in *The Future of Publishing by Scientific and Technical Societies,* 1978, Commission of The European Communities, as cited by Campbell, "Almighty Referee".

[26] For a theoretical discussion of the nature of fraud in science, see War-

ren Schmaus, "Fraud and the Norms of Science", *Science, Technology and Human Values* 8, no. 4 (1983), pp. 12-22; and Harriet Zuckerman, "Norms and Deviant Behaviour in Science", *Science, Technology and Human Values* 9, no. 1 (1984), pp. 7-13.

[27] C. Liébecq, Letter: "Privileged Communication", *Science* 211, (1981), p. 336.

[28] A.S. Relman, Editorial: "Lessons from the Darsee Affair", *The New England Journal of Medicine* 308, no. 23 (1983), pp. 1415-17.

[29] John Irvine, Ben Martin, and Gerry Oldham, Science Policy Research Unit, University of Sussex, U.K., "Research Evaluation in British Science: A SPRU Review", 1983.

[30] S. Ceci, "Science and Deception", transcript of an interview on CBC Radio *Ideas* series, Part IV, p. 1 (1982).

[31] Lewin, "Science and Deception", pp. 1-2.

[32] Peters and Ceci, "Peer-Reveiw Practices of Psychological Journals", p. 187; see also D.P. Peters and S.J. Ceci, "A Manuscript Masquerade", *The Sciences*, September 1980, pp. 16-19, 35.

[33] Stewart and Feder, "The Integrity of the Scientific Literature".

[34] Lazarus, "Interreferee Agreement".

[35] Stephen Cole, Jonathon R. Cole, and Gary A. Simon, "Chance and Consensus in Peer Review", *Science* 214 (1981), p. 881.

[36] For a detailed critique of peer-review procedures, see *Science, Technology and Human Values* 10, no. 52, issue 3 (Summer 1985), special issue on peer review and public policy.

[37] J.A. Rowney and T.J. Zenisek, "Manuscript Characteristics Influencing Reviewers' Decisions", *Canadian Psychology* 21, no. 1 (1980), pp. 17-21.

[38] Alan L. Porter and Frederick A. Rossini, "Peer Review of Interdisciplinary Research Proposals", *Science, Technology and Human Values* 10, no. 52, issue 3 (Summer 1985), pp. 33-38.

[39] M.J. Mahoney, *Scientist as Subject* (Cambridge, Mass.: Ballinger Publishing Company, 1976), p. 93.

[40] C.J. Sindermann, *Winning the Games Scientists Play* (New York: Plenum Press, 1982), p. 23.

[41] S. Harnad, "Peer Commentary on Peer Review", *The Behavioral and Brain Sciences* 5, no. 2 (June 1982), p. 185.

[42] D. Greenberg, "Fraudian Analysis", *New Scientist* 95 (September 2, 1982), p. 643; Allan Franklin, "Forging, Cooking, Trimming and Riding on the Bandwagon", *American Journal of Physics* 52, no. 9 (1984), pp. 786-93.

[43] T.D. Sterling and J.J. Weinbaum, "What Happens When Major Errors Are Discovered Long After an Important Report Has Been Published?", presented at American Statistical Association Annual Meeting, August 1979; J.S. Scott Armstrong, "The Ombudsman: Cheating in Manage-

ment Science", *Interfaces* 13 (1983), pp. 20-29.

[44] Rowney and Zenisek, "Manuscript Characteristics Influencing Reviewers' Decisions", p. 18n.

[45] David F. Horrobin, "Referees and Research Administrators: Barriers to Scientific Research?" *British Medical Journal* 2 (1974), p. 216.

[46] David Horrobin, personal communication, May 1986; Arthur Forer, personal communication, August 1986.

[47] Relman, "Lessons from the Darsee Affair", p. 1417.

[48] D.E Chubin, "Research Malpractice", *BioScience* 35, no. 2 (1985), p. 87.

[49] Rose Sheinin, personal communication, March 9, 1986.

[50] P.F. Ross, "The Sciences' Self Management: Manuscript Refereeing, Peer Review and Goals in Science" (Lincoln, Mass.: The Ross Company, October 1980).

[51] David Horrobin, "Personal View", *British Medical Journal* 4 (November 23, 1974), p. 463; J.M. Tanner, "Referees and Rejects", *British Medical Journal* 2 (May 18, 1974), pp. 381-82.

[52] A.M. Colman, "Editorial Role in Author-Referee Disagreements", *Bulletin of the British Psychological Society* 32 (1979), pp. 390-91.

[53] Chubin, "Research Malpractice".

[54] J. Scott Armstrong, Editorial: "Communication of Research on Forecasting: The Journal", *International Journal of Forecasting* (forthcoming, 1988).

[55] Armstrong, "Cheating in Management Science".

[56] Horrobin, "Referees and Research Administrators"; Tanner, "Referees and Rejects".

[57] Chubin, "Research Malpractice", p. 87.

[58] S. Harnad, personal communication, November 9, 1984.

[59] Ross, "The Sciences Self-Management".

[60] Mahoney, *Scientist as Subject*.

[61] Armstrong, "Communication of Research on Forecasting", and personal communication, November 2, 1987.

[62] I.I. Mitroff, "Designing Peer Review for the Subjective as well as the Objective Side of Science", *The Behavioral and Brain Sciences* 5, no. 2 (June 1982), p. 228.

[63] Rowney and Zenisek, "Manuscript Characteristics Influencing Reviewers' Decisions", p. 20.

[64] D.V. Cicchetti, "On Peer Review: 'We Have Met the Enemy and He Is Us,'" *The Behavioral and Brain Sciences* 5, no. 2 (June 1982), p. 205.

Chapter Seven: Finagling and Funding

[1] Leo Szilard, "The Mark Gable Foundation", in *The Voice of the Dolphins and Other Stories* (New York: Simon and Schuster, 1961), p. 99; David Parnas makes similar allegations about the limitations of research funded by the U.S. Department of Defense in "Star Wars and the Scientific Community", *Alternatives* 13, no. 2 (1986), pp. 27-31.

[2] Rustum Roy, "Alternatives to Review by Peers: A Contribution to the Theory of Scientific Choice", *Minerva* 22 (1984), pp. 316-28.

[3] Ibid.; and see also R. Roy, "Funding Science: The Real Defects of Peer Review and an Alternative to it", *Science, Technology and Human Values* 10, no. 3 (1985), pp. 73-81.

[4] Halvor Aaslestad, "Peer Review at NIH", *Science* 218 (1982), p. 840.

[5] Stephen Cole, Jonathan R. Cole, Gary A. Simon, "Chance and Consensus in Peer Review", *Science* 214 (1981), pp. 881-86.

[6] Stephen Cole, Leonard Rubin, and Jonathon R. Cole, "Peer Review and the Support of Science", *Scientific American* 237 (1977), pp. 40-41.

[7] I.I. Mitroff and D.E. Chubin, "Peer Review at the NSF", *Social Studies of Science* (1979), p. 212, referring to Cole, Rubin, and Cole, "Peer Review and Support of Science", p. 40.

[8] So, too, does Stevan Harnad, although his conclusions are somewhat less critical of NSF peer-review procedures, in "Rational Disagreement in Peer Review", *Science, Technology and Human Values* 10, no. 3 (1985), pp. 55-62.

[9] Mitroff and Chubin, "Peer Review at the NSF", p. 216.

[10] "Scientists Lobby with Success – MRC Budget Increased", University of Toronto *Bulletin*, July 28, 1983, p. 33.

[11] The segments on the MRC investigation and male-only behavioural research are taken from my article, "Who Does Science Serve?" *Canadian Women's Studies* 5, no. 4 (1984), pp. 6-8.

[12] D. Horrobin, "Peer Review: A Philosophically Faulty Concept Which Is Proving Disastrous for Science", *The Behavioral and Brain Sciences* 5, no. 2 (1982), p. 217.

[13] "Research in Peril", *New Scientist*, no. 1471 (August 29, 1985), p. 37.

[14] John Irvine, Ben Martin, and Geoffrey Oldham, "Research Evaluation in British Science: A SPRU Review", Science Policy Research Unit, University of Sussex, U.K., 1983; John Irvine and Ben Martin, "What Direction for Basic Scientific Research?" in M. Gibbons, P. Grimmett,

and B.M. Udgaonkar, eds., *Science and Technology Policy in the 1980s and Beyond* (London: Longmans, 1984).

[15] D. Hensler, "Perceptions of the National Science Foundation Peer Review Process: A Report on a Survey of NSF Reviewers and Applicants", as cited by Mitroff and Chubin, "Peer Review at the NSF.

[16] Alan L. Porter and Frederick A. Rossini, "Peer Review of Interdisciplinary Research Proposals", *Science, Technology and Human Values* 10, no. 3 (1985), p. 34.

[17] E. Borek, "Peering at Peers", *Trends in Biochemical Sciences* 5 (September 1980), pp. 10-11.

[18] M. Kenward, "Peer Review and the Axe Murders", *New Scientist* 102 (31 May 1984), p. 13.

[19] Ibid.

[20] M. Hamer, "No Job, No Government Research Contracts", *New Scientist* 89 (January 21, 1981), pp. 259-60.

[21] J.M. Ziman, in *Public Knowledge* (Cambridge: Cambridge University Press, 1968), pp. 131-32.

[22] David Dickson, *The New Politics of Science* (New York: Pantheon Books, 1984), p. 108, with figures from the U.S. Office of Management and Budget, OMB *Data for Special Analysis K* (Washington, D.C.: OMB, 1983); and Willis H. Shapley et al., *AAAS Report VIII: Research and Development, FY 1984* (Washington, D.C.: American Association for the Advancement of Science, 1983).

[23] C. Stark-Adamec, "Why?" in C. Stark-Adamec, ed., *Sex Roles: Origins, Influences and Implications for Women* (Montreal: Eden Press Women's Publications, 1980), pp. 1-19; C. Stark-Adamec, "Is There a Double Standard in Mental Health Research Findings as well as a Double Standard in Mental Health?" *Ontario Psychologist* 13, no. 3 (1981), pp. 5-16.

[24] Bernard Barber, "Resistance by Scientists to Scientific Discovery", *Science* 134 (1961), pp. 596-602.

[25] M. Planck, *Scientific Autobiography and Other Papers,* translated by E. Gaynor (New York: Philosophical Library, 1949), pp. 33-34.

[26] As reported by M. Friedman in an article by N. Wade, "Why Government Should Not Fund Science", *Science* 210 (1980), p. 33.

[27] C. Stark-Adamec, "Breaking into the Grant Proposal Market", *International Journal of Women's Studies* 4, no. 2 (1981), pp. 105-17; see also D.E. Chubin, G.W. Gillespie, Jr., and G.M. Kurzon, "Funding Success and Failure: Cancer Researchers' Assessments of NIH Peer Review", unpublished manuscript, June 1984.

[28] Several anonymous personal communications.

[29] D.W. Livingstone and R.V. Mason, "Ecological Crisis and the

Autonomy of Science in Capitalist Society", *Alternatives* 8, no. 1 (1978), pp. 3-10, 32; Daryl E. Chubin, "Competence Is Not Enough", *Contemporary Sociology* 9 (1980), pp. 204-7; L. Van Valen and F.A. Pitelka, "Commentary – Intellectual Censorship in Ecology", *Ecology* 55, no. 5 (1974), pp. 925-26; C. Patterson in *Lead in the Human Environment* (Washington: National Academy of Sciences, 1980), pp. 297-98; G.W. Gillespie, Jr., D.E. Chubin, and G.M. Kurzon, "Experience with NIH Peer Review: Researcher's Cynicism and Desire for Change", *Science, Technology and Human Values* 10, no. 3 (1985), pp. 44-54; Chubin, Gillespie, Jr., and Kurzon, "Funding Success and Failure"; Richard A. Muller, "Innovation and Scientific Funding", *Science* 209 (1980), pp. 880-83; D. Greenberg, "Fraudian Analysis", *New Scientist* 95 (September 2, 1982), p. 643; Borek, "Peering at Peers".

[30] D. Chubin and T. Connolly, "Research Trails and Science Policies", *Sociology of the Sciences* 6 (1982), pp. 293-311.

[31] F. Reif and A. Strauss, "The Impact of Rapid Discovery upon the Scientist's Career", *Social Problems* 12, no. 3 (1965), pp. 300-01.

[32] Chubin and Connolly, "Research Trails and Science Policies".

[33] Harvey Brooks, "The Revolution of Technically Intensive Public Policy Disputes", *Science, Technology and Human Values* 9, no. 1 (1984), pp. 39-50.

[34] Quoted in M. Blissett, *Politics in Science* (Boston: Little, Brown & Co., 1972), p. 117.

[35] Robert Rosenthal and Donald B. Rubin, "Interpersonal Expectancy Effects: The First 345 studies", *The Behavioral and Brain Sciences* 3 (1978), pp. 377-415.

[36] R.E. Nisbett and L. Ross, "Human Inference", *Strategies and Shortcomings of Social Judgement* (Englewood Cliffs, N.J.: Prentice-Hall, 1980).

[37] C.A. Anderson, M.R. Lepper, and L. Ross, " The Perseverence of Social Theories: The Role of Explanation in the Persistence of Discredited Information", *Journal of Personality and Social Psychology*, as cited in L. Ross and M.R. Lepper, " The Perseverence of Beliefs: Empirical and Normative Considerations", *New Directions for Methodology of Social and Behavioural Science* 4 (1980), pp. 17-36.

[38] M. Polanyi, "The Potential Theory of Adsorption", *Science* 141, (1963), p. 1013.

[39] Irvine, Martin, and Oldham, "Research Evaluation in British Science"; Irvine and Martin, "What Direction for Basic Scientific Research?"

[40] R. Roy, "An Alternative Funding Mechanism", *Science* 211 (1981), p. 1377; see also Deborah Shapley and Rustum Roy, *Lost at the Frontier* (iSi Press, Philadelphia, 1985).

[41] J. Hackler, "Conflicting Funding Strategies Hurting Research", *Canadian Association of University Teachers Bulletin* (May 1982), pp. 20,

22; F. Mosteller, "Innovation and Evaluation", *Science* 211 (1981), pp. 881-86; R.E. Evenson, P.E. Waggoner, V.W. Ruttan, "Economic Benefits from Research: An Example from Agriculture", *Science* 205 (1979), pp. 1101-07; J.H. Camroe and R.D. Dripps, "Scientific Basis for the Support of Biomedical Science", *Science* 192 (1976), pp. 105-11. For a review of this subject, see John Irvine and Ben R. Martin, *Foresight in Science: Picking the Winners* (London: Frances Pinter, 1984), Chapter 2.

[42] For a discussion of some sectors of the public which have already become involved in science administration and policy development, see Barbara Culliton, "Science's Restive Public", *Daedalus* 107, no. 2 (1978), pp. 147-57.

[43] Dickson, *The New Politics of Science*.

[44] ARMS (Action for Research on Multiple Sclerosis) based in London, England, is an example of an agency funding medical research, where the granting decisions are made entirely by members of the agency who are not professional researchers themselves; examples of research programs involving diverse advisers in an effort to achieve relevance are described in Douglas H. Boucher and Isadore Nabi, "The New World Agriculture Group: A History", *Radical Science Journal* 17 (1985), pp. 88-104; in *Science on Our Side: A New Socialist Agenda for Science, Technology and Medicine* (London: British Society for Social Responsibility and Science, 1982); and in Gabrielle Bammer, Ken Green, and Brian Martin, "Who Gets Kicks Out of Science Policy?" *Search* 17, nos. 1-2 (1986), pp. 41-46.

[45] Porter and Rossini, "Peer Review of Interdisciplinary Research Proposals"; Arthur Forer, personal communication, August 1986.

[46] Alternative analyses and interpretations are encouraged, for example, in the field of social statistics by the authors of the final chapter of *Demystifying Social Statistics*, eds. John Irvine, Ian Miles, and Jeff Evans (London: Pluto Press, 1979).

Conclusions

[1] The debate over criteria for scientific choices was pursued in the pages of *Minerva*, in the 1960s, with articles by: Michael Polanyi, "The Republic of Science" (vol. 1, no. 1 [1962], pp. 54-73); Alvin M. Weinberg, "Criteria for Scientific Choice" (vol. 1, no. 2 [1963], pp. 159-71); Stephen Toulmin, "The Complexity of Scientific Choice: A Stocktaking" (vol. 2, no. 3 [1964], pp. 343-59); and others. It has continued, much more

recently, in *Minerva* and elsewhere, i.e., Jerome R. Ravetz, *Scientific Knowledge and Its Social Problems* (Oxford: Oxford University Press, 1971); John Ziman, "What Are the Options? Social Determinants of Personal Research Plans", *Minerva* 19, no. 1 (1981), pp. 1-42.

[2] William Broad and Nicholas Wade, *Betrayers of the Truth* (New York: Simon and Schuster, 1982).

[3] See, for example, David Joravsky, "Unholy Science", *The New York Review*, October 13, 1983.

[4] Stephen Jay Gould, "Morton's Ranking of Races by Cranial Capacity", *Science* 22 (1978), pp. 503-09.

[5] Allan Franklin, "Forging, Cooking, Trimming, and Riding on the Bandwagon", *American Journal of Physics* 52, no. 9 (1984), pp. 786-93.

[6] Toulmin, "Complexity of Scientific Choice", p. 351.

[7] Erwin Chargaff, "In Praise of Smallness – How Can We Return to Small Science?" *Perspectives in Biology and Medicine* (Spring 1980), pp. 370-86.

[8] Polanyi, "Republic of Science".

[9] C.J. List, "Scientific Fraud: Social Deviance or the Failure of Virtue", *Science, Technology and Human Values* 10, no. 4 (1985), pp. 27-36.

[10] Albert H. Meyerhoff, Senior Attorney, Natural Resources Defense Council, testimony to the Committee on Science and Technology, U.S. House of Representatives, Hearing on University/Industrial Cooperation in Biotechnology, June 16, 1982.

[11] California Rural Legal Assistance represented the California Action Network and fifteen farm workers in a suit against the University of California for failing to represent the public interest.

[12] Witness to this is the establishment, in 1982, of the Canadian Social Sciences and Humanities Research Council's Woman and Work Program.

[13] Karen Messing, Co-Director, Groupe de Recherche-Action en Biologie du Travail, personal communication, November 6, 1987; Donna Mergler, "Worker Participation in Occupational Health Research: Theory and Practice", *International Journal of Health Services* 17, no. 1 (1987).

[14] T.C. Chamberlain, "The Method of Multiple Working Hypotheses", *Science* 148 (1965), pp. 754-59.

[15] The journal *The Behavioral and Brain Sciences*, edited by Stevan Harnad and published by Cambridge University Press, has adopted an unusual format in which an entire issue is devoted to the criticism and defence of one provocative article.

[16] The journal *Medical Hypotheses*, edited by David Horrobin, does this, as does, to a lesser extent, *International Journal of Forecasting*, edited by J. Scott Armstrong and Robert Fildes.

[17] John B. Robinson, "Unlearning and Backcasting: Rethinking Some of the Questions We Ask About the Future", *Technological Forecasting and Social Change* (forthcoming, 1987).

[18] The impact of such measures on general insurance premiums is, of course, a separate issue. Monitoring of rates and their relationship to laboratory purity protocols would, however, have to be incorporated into new laboratory insurance schemes.

INDEX